Professor Eric Laithwaite was born in 1921 and spent his school days in the agricultural Fylde district of Lancashire. He served in the Royal Air Force from 1941 to 1946, rising to the rank of Flying Officer. He did experimental flying at the Royal Aircraft Establishment at Farnborough.

After the war he went to Manchester University to read Electrical Engineering, stayed on to do a Masters degree, working on the world's first commercial, full-scale digital computer. In 1952 he began studying linear induction motors, a subject which occupied him for the next 40 years. Manchester University rewarded him with a PhD and later a DSc.

He made many inventions in linear motors and in 1966 was awarded the Royal Society's S.G. Brown Medal for invention. Subsequently he received the Nikola Tesla Medal of the American Institution of Electrical and Electronic Engineers for further work on the same subject. He was made a Fellow of Imperial College in 1991, and an Honorary Fellow of the Institution of Electrical Engineers in 1992. His linear motor work has taken him to Hong Kong, Florida, California and Toronto, as well as to Zagreb and other European countries.

Professor Laithwaite has made many television appearances, but none more memorable than his presentation of the famous Christmas Lecture series from the Royal Institution in 1966, and again in 1973. In the latter series he included work on gyroscopes, which became the subject of much criticism and debate. But the biology he learned at school never left him, and he has lectured on engineering and biology for over 20 years.

He is now an Emeritus Professor of Imperial College and a Visiting Professor of The University of Sussex, where he continues his research on gyroscopes, because he has seen unusual phenomena 'that won't go away!' He is married with four children and three grandchildren.

So many of our most clever inventions have been anticipated by nature, often by millions of years. Professor Laithwaite is an inventor and engineer fascinated by the wonderful genius of nature. The author writes beautifully, allowing us to share in his infectious excitement. Through humour, anecdote and science, the reader is presented with a hugely enjoyable inventor's journey through the world of nature.

The shapes and sizes of creatures and machines alike have been largely dictated by the size and weight of the earth and by the properties of materials. On the way to arriving at these conclusions the author crosses many boundaries between engineering, biology, religion, economics and cosmology, but all in the lay-person's terms.

This is a book for all thinking people with or without a scientific background.

Related titles

Remarkable Discoveries!
FRANK ASHALL
Hard to Swallow: a brief history of food
RICHARD LACEY
The Outer Reaches of Life
JOHN POSTGATE
Prometheus Bound:
science in dynamic steady state
JOHN ZIMAN

ERIC LAITHWAITE

An Inventor in the Garden of Eden

CAMBRIDGE
UNIVERSITY PRESS

Published by the Press Syndicate of the University of Cambridge
The Pitt Building, Trumpington Street, Cambridge CB2 1RP
40 West 20th Street, New York, NY 10011-4211, USA
10 Stamford Road, Oakleigh, Melbourne 3166, Australia

First published 1994

Printed in Great Britain by
Biddles Ltd, Guildford

A catalogue record for this book is available from the British Library

Library of Congress cataloguing in publication data

Laithwaite, E. R. (Eric Roberts)
An inventor in the Garden of Eden / Eric Laithwaite.
p.   cm.
Includes bibliographical references and index.
ISBN 0 521 44106 4
1. Laithwaite, E. R. (Eric Roberts)   2. Inventors – Great Britain –
– Biography.   I. Title.
T40.L35L35   1994
609.2 – dc20
[B] 94-7459 CIP

ISBN 0 521 44106 4 hardback

To Bernard Stevenson

Contents

Preface

Eighteen years at Manchester University and a further 22 at Imperial College, London, would appear to qualify me as a city-dweller. Yet I have always thought of myself as a countryman, spending my formative years in the essentially agricultural district of Fylde in Lancashire.

I entered Grammar School at the age of 10 and was introduced to 'new' subjects such as French, Latin, physics and biology. In physics, they taught me about force, work and energy and I was told that if I picked up a heavy sack of potatoes from one corner of a room, carried it to the opposite corner and put it down, I had done no work. This seemed to contradict 'common sense', but there it was – that was 'Physics'!

Biology I enjoyed a great deal, not least because the biology master, Bernard Stevenson, was such an inspiration to us all. As well as making his subject fascinating, he was every schoolboy's idea of the ideal man. For his inspiration, which for me has lasted over 60 years, I dedicate this book to him.

By the time I was 12, I was told that I now had to choose between physics and biology since the lessons in these subjects had to be conducted at the same times 'because of time-tabling and staffing difficulties' (I'll bet it's much the same today in many schools!). We were given about 10 minutes to choose in what was perhaps one of the most influential cross-roads in our lives. I felt like shouting: 'But I want to do BOTH!', but I knew it would be useless.

So I argued with myself along these lines: 'Physics seems to be mostly sums, biology mostly essays. I like biology better than physics, but I'm better at sums than I am at essays. Anyway, my best friend is going to do biology, so I can keep asking him about it and keep in touch that way. That does it – I'll do physics.'

In the stage musical *Kismet*, Oscar Hammerstein wrote: 'Is fortune predicated on such tiny terms as these?' Indeed, where might I have been today had I chosen biology at the age of 12? As it was, I left school to be called up, and spent the war years in the RAF where I 'discovered' engineering, went to Manchester University to study it in 1946 and got involved in this very, very new subject of 'computers', in particular being secretary to the inaugural conference on the world's first full-scale commercial machine (the Ferranti Mark I) in 1951. I then returned to 'heavy' engineering, went on to specialise in induction motors and, more particularly, in *linear* induction motors.

But by 1977 the Institution of Electrical Engineers had made me Chairman of the Power Board (there were four Boards in those days) and I was due to give my Chairman's Address. Looking for something new to talk about, I realised that almost everything I had done in my induction motor research had been basically about *shape* and about *changing* the shapes of machines. Yes, of course, I'll tell them about shape. So my address was called 'The shape of things to come' – not an original title but it expressed the sentiment about machines research adequately.

This is really where the book began, because my ever-present interest in biology, and in butterflies in particular, made me introduce some comparisons between engineering products and living things. The audience enjoyed it and I was asked to give it again at different centres up and down the country.

Now, I dislike giving the same lecture twice. It is inclined to bore me the second time and if I get bored I am liable to bore the audience too. So I varied it, built up a stock of some 300 colour slides from which I could make a selection each time and the range of titles ran through:

> Engineering and Nature Study
> Man-made, God-made
> The Quick and the Dead
> Insects and Innovation

and many other similar titles. In all, I gave over 100 lectures on this theme and rounded them off with a programme on BBC *Horizon* in 1989 called *Gaze in Wonder*, so by the time I came to put pen to paper for this book I had effectively spent some 25 years thinking about it.

In 1966 the Royal Society awarded me the S. G. Brown Medal for Invention, and since I have no formal training in biology, I tend to view the natural world in the same naive way that our ultimate ancestors must have viewed Eden – hence the title.

Eric R. Laithwaite

Acknowledgements

I am most grateful to Robert Harington for his continued encouragement and help during the preparation of this book. My thanks are also due to my friend and colleague Bill Dawson, who read the original manuscript and made many useful suggestions. The manuscript was kindly typed by Elaine Taylor, to whom also, my thanks.

Illustration credits

Figure 2.5 comes from *Ancient Engineers* by L. Sprague de Camp. Tandem (London), 1917.

Figure 4.1 is courtesy of Tony Pacey © 1991.

Figure 4.2 is from *Grays's Anatomy*, 35th edition, edited by R. Warwick and P. Williams. Longman (London), 1973.

Figure 4.9 is from *Hildebrand's Travel Guide – Seychelles*, distributed in U.K. by Harrop Columbus (London), 1990.

Figure 4.6 is from *Growth and Form* by D'Arcy Thompson. Cambridge University Press (Cambridge), 1961.

Figures 4.17 to 4.19 are from *Magnifications* by David Scharf. Frederick Muller Ltd (London), 1977.

Figure 6.8(d), 6.17 to 6.22 and 7.12(c) are from *Patterns in Nature* by Peter Stevens. Peregrine Books (London), 1976.

Figure 8.6 is from *Nature, Mother of Invention* by Felix Paturi. Thames and Hudson (London), 1976.

1

Gardens of Eden

Any photograph or actual view of a suburban scene will suffice to pose the simple question: 'Can you tell, at a glance, which objects that you see were made by human hands and which are alive and growing?' The answer is obviously 'Yes'. But the slightly more difficult question is: '*How* did you decide?' A first answer might be: 'By familiarity. I have seen many similar houses, and vehicles, and trees.' But is there, perhaps, one single rule that would guide you in a situation where the objects that were seen were less familiar? The best answer you will find to this question is surely 'shape' and, in particular, 'regularity'. Certainly colour is not a criterion. We can imitate all of nature's colours on canvas. We can outshine the most brilliant by the use of fluorescent paints. It is the shapes that are far more decisive. The shapes we make are still dominated by Euclidian geometry. We base so many of our creations on straight lines and circles, on cubes and brick shapes, on pyramids, cylinders, spheres and right circular cones – and we do most of it in the name of 'economics'. The shapes of nature we see as over-elaborate, wasteful and unnecessary. We see nature as having missed out on the wheel and as being incapable of using pure metals.

We have long seen humans as the superior beings on this planet, given 'dominion' over the lower animals, given the earth to 'subdue it'. It is impossible even to begin a treatise on the relationship between the Man-made and the God-made without involving religion, questioning evolution and even touching such subjects as extrasensory perception, students of which subject are still regarded by many 'pure' scientists as hardly respectable: topology for sure, will be our staple diet. Science will take second place to engineering and to the other technologies. This is no mere collection of facts. The text will pose a thousand questions for every one answered. But, hopefully, profitable messages will emerge out of the mist, however vast that which is obscure. One has only to contemplate the size of *Gray's Anatomy*[1] to realise that its contents relate to but one single species of mammal. Then to realise that there are over 400 000 species of beetle, nearly 200 000 species of butterfly and moth, is to be appalled at what one has undertaken to attempt.

Many facets of the subject and many profitable comparisons between the manufactured and the living will doubtless have been overlooked. A nature book on the same theme with the title *Nature, Mother of Invention*[2] was written by a botanist, and the opportunities that he missed only served to daunt the more in attempting the same task seen from 'the other side' (the engineering side). One is also mindful of a quotation from Max Delbrück, Nobel prizewinner for physiology: 'When a physicist or engineer starts work on a biological problem he is always afraid that he will not know enough biology. It invariably emerges that he did not know enough physics or engineering.'

Where to begin

Let us start with a little heart-searching that is calculated to bring a stream of humility before the end of the text. We humans would claim to be three-dimensional beings, whatever else we are. We are conscious that objects have length, breadth and depth. When we study topics of the nature of electromagnetism, or when we reach out for knowledge of the almost infinitely large (the Universe) or the almost infinitely small (the 'fundamental particle'), we are vaguely conscious of a fourth dimension that we cannot fully appreciate. Yet as our civilisation progresses we become more and more chained to two-dimensional objects which often require a 'translation' to 3-D inside the brain. The great majority of our learning is through the media of flat sheets of paper (book pages), flat chalkboards and cinema screens and the flat faces of

TV tubes. We have become so skilled at 'seeing' a 2-D picture as a 3-D object that the 'program' (to use computer language) has been transferred from the consciously thinking part of our brain to the automatic part which governs so many of our actions, such as walking, eating, picking up and handling objects, and so on. Once this transfer has been made it is easy to alarm the brain by relatively simple tricks, and the sight of the 'object' shown in Fig. 1.1 is most disturbing, especially to engineers. This drawing makes a good starting point for the many discussions of topology which we shall make in establishing profitable comparisons between the shapes of nature and those made by our own hands. Let us take Fig. 1.1 for what it is, and no more: lines on a piece of paper, some of which form closed areas whilst others merely

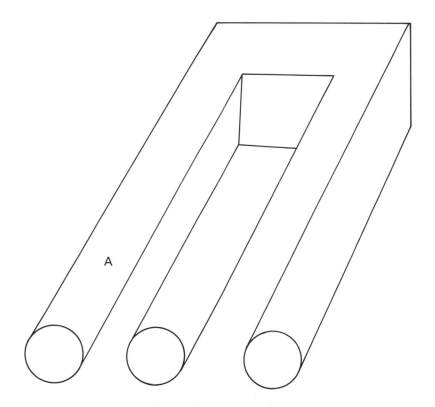

Fig. 1.1. An impossible object.

join such areas together. Figure 1.2 is topologically identical to
Fig. 1.1, and no-one is going to see *this* as three-dimensional.
What is interesting is the extent to which we become conditioned
to the 2-D–3-D 'translations'. When we are young, Fig. 1.1
seems almost as 'flat' as Fig. 1.2. Given crayons to colour it, we
will fill in all the fully bounded areas as shown in Fig. 1.3, making
no attempt to start colouring the space A, which we see at once
is not totally contained by lines. It is also interesting to show how
Fig. 1.1 loses its '3-D-ness' when coloured, and which combination of colours gives the greatest sensation of flatness. Readers
who are 'of riper years' might like to test their visual memory by
trying to re-draw Fig. 1.1 some 24 hours after seeing it in these
pages. It is not half so easy as it appears.

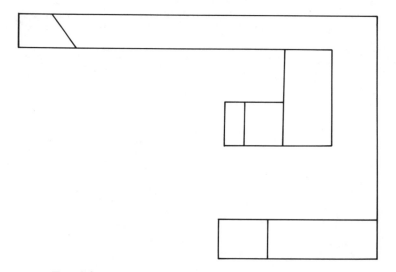

Fig. 1.2. A drawing topologically the same as Fig. 1.1.

Evolution and design

Quite apart from considerations of wheels and pure metals, many
readers might be prepared to go along with the idea that the
greatest dividing wall between manufactured articles and living

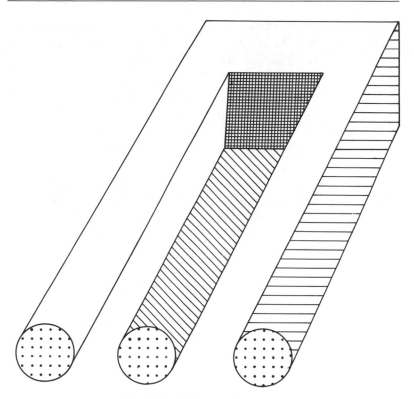

Fig. 1.3. Filling in the spaces.

organisms is that the former are the result of 'design', whilst the latter are products of 'evolution'. Let me say at once that the devout will always maintain that our 'creator' designed all things. If that be true, He, She or It made some monumental mistakes in the process – like giving wings to ostriches. What is much more certain is that remarkably few of the products of modern technology have in fact been *designed*. By this I mean that the person who thinks about the layout of the parts of a motor car for next year-but-one's Motor Show is much more conscious of what will appeal to the customer and what will make more profit than of what will actually work better. The 'designer' is going to make use of as much of the experience of all earlier designers as possible. After all, is there a better process for rapid development than learning by the mistakes of others?

Let me illustrate what I mean by using a much simpler manufactured article – the common teapot (Fig. 1.4). If all teapots had
the date of manufacture stamped on them under the base, it
would be impossible to guess the date on this one to within 50
years or more. This 'design' might be compared, on a vastly
reduced time scale, to the common woodlouse that has remained
unchanged by the ravages of this planet and by the processes of
evolution for millions of years. We might conclude that both
woodlouse and teapot are examples of 'near-perfect' designs.

Let us, however, look at the teapot as might a fifth form school
pupil doing a scientific experiment. Such pupils are taught to
write formally under headings such as 'Purpose', and in this case
what would follow would be: 'A vessel for the brewing and dispen-
sing of a liquid which is called "tea".' So we will examine it as
an instrument for this purpose. The brewing facility is fine. The
pot does not crack when subjected to a sudden influx of water at
100 °C. A little stirring done, we begin the process of pouring.
We are surprised by the weight of a full pot and since the handle
is on one side, the centre of mass is several inches displaced from
the centre of grip. It tips and we pour a little tea before we
intended – usually on to the tablecloth. At this point we must ask

Fig. 1.4. A traditional teapot.

why the handle was so placed. Could it be because there is a small hole in the lid to prevent steam from building up pressure inside and rattling the lid? Steam escaping from the hole would scald our hand if the handle were situated over the mass centre. True, but we have had experience with kettles like the one shown in Fig. 1.5, a typical modern 'design' with a vent in the lid so placed in relation to the handle that the designer was obviously intent on scalding everyone who used it! Kettles, if anything, get hotter than teapots. Yet the usual designs invariably have the handle on top.

Fig. 1.5. A modern kettle.

Let us continue the experiment by observing now that perhaps the design *did* incorporate a device to be used in the eventuality of the pourer finding the force of gravity surprisingly large. The tip of the spout is curved downwards so as to present a hook-shaped profile below, into which the forefinger of the free hand can be placed to balance the weight. However such an eventuality is *bound* to result in a scalded forefinger. Perhaps it is not so. Perhaps the obviously elegant shape of the entire spout has been designed to produce a beautifully smooth stream of tea, 'laminar flow' as the engineer calls it. We pour. It is a near disaster. The tea splatters out in several directions, some goes in the cup, some in the saucer, a little on the tablecloth and who knows, with some luck, a little on the person for whom the cup was intended. 'Ah', they say, 'it always pours badly when it is nearly full'. Generally it also pours badly when it is nearly empty. You decide that as soon as shops re-open tomorrow morning you will go and buy an attachment for that spout that acts like a nozzle on a water tap.

You reflect on why the designer did not incorporate the properties of the attachment into the spout design, and come to the conclusion that it might have looked ugly; the attachment *certainly* looks ugly. That does not seem to matter; at least let us say that you did not contemplate the possibility that you were acquiring a 'rogue pourer' when you bought it. (But when you finally smash the old brown pot you will go and buy another just like it. And you will not ask for a pouring demonstration before purchase. Such things are simply 'not done'.) Could it be that the teapot designer deliberately built in a 'turbulator' (my own word) to ensure a steady sale for the attachment – perish the thought!

We have now reached the stage where we have poured several cups and the quality of the pouring has undoubtedly improved. What *was* the basic cause of the turbulence? Across the entrance to the spout there is a baffle or filter in the form of a wall containing holes. What is *that* particular feature for?

> As everyone knows, it is to stop tea leaves from coming out with the tea.
>
> Have you seen the size of the holes?
>
> The modern tea leaves could emerge broadside on, twenty abreast!
>
> But of course, most tea addicts of today don't use raw tea leaves; they use teabags.
>
> It'll stop teabags.
>
> The only trouble is that as soon as it has arrested the first teabag, it stops the tea as well!

Why does the modern 'Old Brown' teapot still retain this strange feature? The answer is fairly obvious if you look inside the pot. Imagine yourself as the potter with his lump of clay. He is going to make the body of the pot first and stick the spout and handle on later. To make a hole the size of the base of the spout and later to join it with a spout leaving a smooth finish when you can only get at the outside in the process is difficult. What is far easier is to poke a few holes in a cluster, smooth off the surface and simply stick the spout in place – simple economics.

Back to our tea pourer. When the pot is nearly empty the lid falls off, breaking the cup, and flooding the table with tea. As a piece of scientific equipment the Old Brown is a disaster.

Has nobody ever done better? Well, yes. In 1972 Rosenthal of Germany produced what was perhaps the first teapot ever to be designed scientifically (Fig. 1.6). See at once how elegant it is, constituting a superb example of the philosophy that things that *look* good, *are* good (about which, more later). Without handling it you can see that the centre of mass of pot and tea will lie directly under the first finger at all times, no matter what the pouring angle nor how full it is. It was *designed* to be like that. No spout entrance baffle to act as turbulator, it pours with laminar flow whether you allow a mere trickle of the beloved liquid, or whether you heave it straight over at 90°. It is an interesting experiment to fill this pot with water, give it to someone who has

Fig. 1.6. A teapot designed by an engineer.

never handled it and ask them to test its pouring qualities by suddenly rotating it through 90° over a sink or bucket. Nineteen people out of 20 will immediately put their thumb on the lid. It is *tradition* to do so. It has even become, for most of us, instinctive. But the lid does not fall off until you have rotated the pot from its normal position through 135°, which is a silly way to pour tea! It was *designed* to eliminate lid fall.

I have heard it criticised 'because you can't clean the inside'. I am certain that if the restless human mind can devise such a beautiful and functional implement, it can devise a brush to clean it. (Connoisseurs of tea-making tell me you never clean the insides of pots anyway.)

Now, the scientific pot was expensive (£13.95 in 1972). Since when has price had anything to do with popularity? People have birthdays and anniversaries and then there is Christmas. There was a steady trade in Mah Jong sets made from fine split cane and ivory (before ivory was rightly frowned upon), an even bigger trade in expensive chess sets. How much more readily should an article whose use always brings pleasure in a creature-comfort sort of way, three or four times a day, be snapped up? The proof of the pudding was in the eating, of course. How many of these teapots have you seen in the last 5 or 6 years? They hardly sold at all! The design was discontinued at an early date.

I discovered that it had been designed by an Italian engineer who was better known for his design of racing cars. He was a fluid flow specialist. The problem of pouring liquid at 1 mile per hour is similar to that of forcing an object through a gas (air) at 200 mph. Now, why was it not popular? Figure 1.6 shows you its best profile. Seen in plan it is at once obvious that its shape is equally functional. Apart from the spout, it is circular. A circle has maximum area for a given perimeter. It is an 'economic' shape. But it looks like a bed-pan!

The Old Brown teapot is, if only subconsciously, one of your earliest memories, perhaps of visiting Auntie or Granny on Sundays, when you were 5. It was always a happy time and there was always an Old Brown at tea-time.

So, what have we learned from our study of teapots? Undoubtedly something about tradition. Now, in many facets of engineering manufacture, 'tradition' is an excuse for leaving things as they are – for laziness, if you like. No, it is more than that. The first rule of production engineering which reflects right back into articles on the shop counter is that the more you make and sell, the cheaper the price per article. The reason is fairly obvious. No new tools are needed after the first one comes off the production line. The more you make, the better the shop floor operations become. There are fewer rejects, the speed of manufacture increases. Less concentration is required by the operatives; they get production bonuses. They are happier people. It all makes good sense to the economist.

But tradition can be the death of mighty industries. Of this there is no more dramatic example than that of the Lancashire cotton industry. The Lancashire loom was, in the 1930s, along with the umbrella, the best bargain in engineering that could be bought. (Even in the inflated 1960s, an automatically opening umbrella cost only £1.) For £400 you could buy a loom that would weave cloth 54 inches wide, putting in weft threads at more than one a second. The Lancashire loom was an example of an invention 50 years ahead of its time that continued in use until it was 50 years *behind* its time. The philosophy 'What was good enough for my father is good enough for me' was rife in Lancashire in the 1930s where the self-taught directors of industry *boasted* of 'not having been to a university'. Argue with them, as I did, about the possibilities of incorporating linear electric motors for shuttle propulsion and you were told 'It can't go much faster safely', 'Noise doesn't matter' and similar false beliefs. No-one seemed to notice that the Swiss were making looms *without* shuttles, that ran as smoothly as Swiss watches, nor that the textile industry was swinging over from weaving to knitting as the result of tremendous advances in automatic knitting machines. There were a few prophets, one or two of whom I was privileged to know, but no-one wanted to hear them. The Lancashire 'tackler' (who used to take the place of the Irishman as the butt of all the wit of comedians in the UK), whose job was simply to keep looms running, was a skilful man and his principal tools were a spokeshave and a torn up cigarette packet (used frequently as spacing washer or whatever). But his days were numbered, and he never

knew until it was too late. Tradition, pure tradition, engulfed him
and his fellow operatives. He died like a dinosaur.

So we must arrive at the conclusion that the process of design
is, at best, a combination of both our conscious and our uncon-
scious mind, which, having undoubtedly created new designs in
our time, demands that everything in nature was similarly
designed with purpose aforethought – and it need not be so.

2
The human thinker

The only thinker?

In my student days I lived in a students' hall of residence which was a Church of England establishment. It had a small chapel in which daily devotions were carried out and it was customary for students of theology to conduct some of the services. One evening one of my friends (a theologian) began a prayer with the words 'Let us pray for our friends the animals.' After the service I tackled him on the subject of whether an animal has a soul. He said that to the extent that certain animal species had come to live closer and closer to us, to share our houses, our cooked food and many of our habits, they have at least the beginnings of a soul.

The same, I am sure, may be said of the process of thinking. During the Royal Institute Christmas Lectures of 1974[1] I used a film clip kindly loaned by the BBC. It showed the antics of a male chimpanzee, seeing himself in a mirror for the first time. After a period of observing that the 'other chap' copied his every movement, he went round the back of the mirror to meet him. The non-event caused him to leap back in front of the mirror – that other fellow was still there! He repeated the exercise three times in all. Then he quite clearly *thought* of a better idea. Moving close to the mirror, and hence to the 'other' chimp, he put his arm around the back of the mirror and *felt* for him. This negative result was more than he could bear, alone. So off he went to fetch his friends to show them what he had discovered.

If one argues that all of these actions were the result of habit-forming, without *conscious* thought, then we cast doubt at once on our own thinking ability. Personally I would prefer to go along with the idea that thought is more than a reflex and that other animals are capable of it to a degree.

Habit

It is necessary to proceed with some caution, however, before accrediting all our actions to conscious thought, or even the idea that the brain, which 'stores' (a computer word) subroutines such as how to walk, turn the head, use the fingers, and so on, leaves the thinking part (with all its inadequacies) free to grapple with new, or at least less familiar, requirements.

Under the heading of 'habit' we should include conditional reflex. By such means, wild animals may be 'trained' to perform most of the tricks we see in circus acts,* right down to the cage bird in the home which has 'learned' to pull a chain with its beak, hold it down with its foot, and repeat the process until a small bucket containing food or water comes within reach.

Experiments whose results are at first more disturbing to most of us but which give a much clearer definition of habit, can be carried out on the 'lower orders' of creatures: worms, insects, etc. These experiments go a long way in helping us to recognise the parts of ourselves that are *animal*, in the most basic sense.

If fish are taken from a lake and kept for years in a square garden pond, measuring 4 feet by 4 feet in plan, at first they will be afraid of the approach of human footsteps, flee in panic and bang their noses on the concrete vertical sides of the pond. After a few days they all have very sore noses. But they soon 'learn' the geometry of the pond, and in a panic they dart around the perimeter, stopping short of collision with the wall by less than a quarter of an inch each time. They have become 'habit-formed'. The process hinges essentially on cause and effect – 'If I do that, I shall get hurt.' The phenomenon is by no means restricted to

---

* I do not include dolphins under the heading 'wild'.

a punishment for effect. The cage bird referred to earlier got a *reward*. Most circus animals (including dolphins) are 'trained' by the reward-ending effect, whereas the legendary eskimo 'trained' a dog team largely by punishment. *Homo sapiens* uses both in bringing up children to conform to society and becomes horribly tangled up in arguments about child psychology, about what is right and what is wrong, and the children suffer possibly even more than do the eskimo's dogs, and some, hardly surprisingly, opt out of 'the system' altogether.

But let us return to our fish in their 4 foot by 4 foot world. After 2 years we return them to the lake, and for half an hour they swim in a square path, just the size of the pond, less half an inch in each direction. No amount of jumping about on the bank, even of throwing stones into their little world, will make them penetrate the mental walls of concrete that habit has created. They resemble a school pupil who has been taught conventional physics in the belief that this is all there is in the universe!

The phenomenon of habit goes very deep into the brain. It can take only a few weeks to acquire a habit, good or bad, and it can then be retained for a lifetime. A habit can easily become confused with 'instinct'. It has nothing to do with hypnosis or the effects of artificial introduction of powerful drugs. It is part of the animal that is in all of us.

I first encountered it while studying caterpillars, when I was still at school. The tiny creatures (about half an inch long only) were gregarious and feeding ravenously on my father's vegetables. I discovered that if I clapped my hands, each caterpillar would flick its tail in the air, the purpose undoubtedly being to flick off an approaching parasitic fly, intent on laying an egg under the skin of its caterpillar victim. The sound of the fly's wings vibrating would be at a frequency of several kilohertz (kHz). But a hand clap contains most frequencies between 10 and 100 000 Hz, so it contains the relevant frequency in this case. I wondered if they would ever tire of repeating the 'physical jerk' if I clapped every 2 seconds, for example. Doubtless there would come a time when exhaustion overcame the reflex. In this case it overcame me (or rather I ran out of patience) before I ever got near the point when they 'learned' to ignore the sound. What really shook me was that when I finally stopped, they all did at least two extra flicks, some three, some four, all in the strict rhythm I had taught them – a 2-second interval between flicks. They had been habit-formed in a subroutine in but a few minutes. It was many years later that I argued that perhaps they had become conscious of 'negative sound' – a concept comparable with a 'positive hole' in a semiconductor. They expected a sound every 2 seconds and when it failed to arrive they responded to the *absence* of it? – I think not, but the idea was good!

Experience

In addition to confusing habit with learning and with instinct, it is almost inseparable from the effect of the passage of time which we have called 'experience'. There is a vast difference, however, between the explorations of a small baby, once it discovers that there are other things besides its mother, and the built-in tail flick of the caterpillar which can be taught a rhythm habit in the space of minutes. One of the delights of a parent is to see the pleasure that is brought to its young offspring by the latter's discovery of gravity. It reaches a bag of tomatoes on its perambulator cover, holds one in its tiny hand which overhangs the side and lets go. Three seconds of pure glee and a second tomato follows the first, and another and another . . . ! The baby is teaching itself its first *craft*, and is doing so in the best traditions of the great scientists and technologists – by experiment.

Experience does not even have to be sought. It happens to everyone whether they want it or not. A wise old man once said to me: 'I never met a man who couldn't teach me something,' and then added quickly: 'And I never met a man I couldn't teach something.' The important thing, of course, is the sifting of the more profitable experiences from the less profitable ones, and in this it resembles the aims of the engineer.

Experience, of course, depends on memory. No-one is bereft of all memory and no-one ever has enough of it. It has been said that the human brain automatically stores the registration numbers of all motor vehicles that the eyes have seen, provided the 'seeing' involved consciously registering those numbers. Long-forgotten facts can be brought back under hypnosis or through the use of drugs.

Logic

The place where many would divide the human from the animal is in matters of cause and effect that can be predicted. The fact that the sight of a lioness will put a herd of wildebeest to flight is typical of what the majority of us would classify as 'instinct' on the part of the wildebeest, rather than a conclusion made on the basis of logic. But how they cope with the chimpanzee's behaviour in relation to the mirror I have no idea. Instinct will have to be stretched to unacceptable lengths to take such phenomena under its umbrella. In fact, we may profitably place such phenomena in a class of their own which could be headed 'adaptability'. This includes the behaviour of living things in situations of which they had no previous knowledge, but more, in situations for which they had never been designed or orientated by the process of evolution. This important topic is worthy of a fuller treatment than is fitting in a paragraph on logic. Readers may indeed care to turn to pp. 238–242 at this point and read the paragraph on adaptation now, before proceeding in pursuit of logic.

Up to the turn of the last century, no hypothesis in science was acceptable unless it was seen to be logical, for in logic all things are to be locked to each other as perfectly as are the pieces of a completed jigsaw. They are a part of what was once a whole thing, and science was seen as a continuing struggle to remake that 'thing'. Whichever path you choose to complete a jigsaw, none will ever preclude fitting in the rest of the pieces. If you begin, as do many, by finding all the pieces with a flat edge and completing the border first, then it is obvious that it is of no consequence where you start. If the pieces of the border were to be numbered 1 to 113, where no. 113 links with no. 1, then you may begin with piece 57 and join it to no. 58, then to 59 and so

on, or you may join 57 to 56 and proceed the other way round.
Science was like that.

By the time it was believed that all matter consisted of protons,
neutrons and electrons in combination, the jigsaw seemed to be
complete, and with that belief came perfect atheism, for the laws
of gravity, of magnetism and of electrostatics seemed adequate
to predict the future of all things in the universe, provided one
had enough knowledge of the position and motion of all the 'bits'
at any one time. So there was no such thing as free will. Everything was predestined, and not necessarily by a Supreme Being
– only the cold logic of three equations, involving the fundamental
'bits' of mass, charge and magnetic pole. Some doubted but few
questioned these equations that spoke with ultimate *authority*.
There was no need to define things like mass beyond such obvious
statements as 'mass is the amount of matter in a body'.

Sixty years on and the fault lines in this 'logical' ideology are
beginning to show. A few examples are listed below.

1. The word 'amount' in the definition of mass has no mean-
    ing. It is certainly not the volume or we would have called
    it 'volume'. Yet even with our knowledge of fundamental
    particles, you do not count the particles to arrive at a
    meaning for 'amount'. Let's face it, it *had* no meaning.

2. We are in an even worse situation if we attempt to define
    'charge'. 'It is a thing with no mass that produces force
    on all other things like itself'. Try explaining charge to
    an intelligent child of 12 without using the word 'thing'.

3. The 'arbitrary' constants in two of the three equations
    were found to be connected (through relativity theory),
    whereas neither appeared to be connected to that in the
    third equation.

Einstein declared the connection between electromagnetism
and a 'gravito-inertial' system to be 'the key to the cosmos',
although neither he nor anyone else has ever found it.

Came Heisenberg, Planck, Bohr, Pauli and others and a new
science emerged, so revolutionary that it left the person in the
street pawing in the air. *It had abandoned logic!*

Under the conventional laws of mechanics a pendulum bob swinging freely between given limits is more likely to be found near the ends of its swing, at any given time, than it is near the middle. This is because it is going more slowly near the ends and therefore spends a longer time there than it does near the centre. But under the new laws of mechanics (quantum mechanics) it has a finite chance of being found outside its amplitude altogether. Has the world gone mad? These were the thoughts and arguments that raged over the first half of the twentieth century, but in the end the 'new guard' emerged in triumph. Logic was doomed in a study of the *real* world! The future was as unpredictable as ever and, what is more, could now be *proved* to be so.

It is only in the last few decades that the ideas expressed by quantum mechanics and by relativity (curved space, a fourth dimension, etc.) have lost their mystique and, as usual, this was because the young were prepared to believe what their elders and betters were not, and *they* (the young) became the new teachers with their feet on a new ground. For me there is no better summary of the decline and fall of logic than the words of Sir Arthur Eddington.[2]

> However successful the theory of a four-dimensional world may be, it is difficult to ignore a voice inside us which whispers 'At the back of your mind, you know that a fourth dimension is all nonsense'. I fancy that voice must often have had a busy time in the past history of physics. What nonsense to say that this solid table on which I am writing is a collection of electrons moving with prodigious speed in empty spaces, which relatively to electronic dimensions are as wide as the spaces between the planets in the solar system! What nonsense to say that the thin air is trying to crush my body with a load of 14 lb to the square inch! What nonsense that the star cluster which I see through the telescope, obviously there *now*, is a glimpse into a past age 50,000 years ago! Let us not be beguiled by this voice. It is discredited . . .
>
> We have found a strange footprint on the shores of the unknown. We have devised profound theories, one after

another to account for its origin. At last, we have succeeded
in reconstructing the creature that made the footprint. And
lo! It is our own.

We have dug the bear pits in science and then fallen in them
ourselves.

An alien being, hundreds of times the size of a human (linear
dimensions), living on a planet millions of times the volume of
the sun, but having a density no more than that of a gas, has
invented a telescope so powerful that it can focus on a tiny part
of the Universe so small that it may consist of a square mile of
the earth's surface only. The alien, who calls a telescope an
'electron microscope', has it focused on a furniture factory on
earth. The creature sees trucks carrying tree trunks into the factory, and because it is so large its time scale is such as to make
the tree trunks appear to be fired into the factory (the 'nucleus',
in alien speak) at very high speeds. From the exits of the factory
come loads of chairs, moving equally fast. Correctly, the alien
deduces that the injection of logs was the direct cause of the
outgoing of chairs. Incorrectly, it deduces that the factory is made
up of chairs. The analogy with particle physics is not too obscure.

Simplicity

As thinkers, we have long prided ourselves on having 'tidy' minds.
Curiosity, which is by no means limited to the *sapiens* species,
led us to search for the ultimate in time and space, i.e. length.
We have probed the Universe for more and more distant things.
We have explored the smaller and smaller in the hope of finding
that all matter consists of only a few basic bits. In the fruitless
hope of understanding 'how it all worked', our science led us to
simplify the complex to get a 'feel' for the problems. Our masses
shrunk to 'point masses', a point having no length, rather it merely
marked a position. A line had no width, a surface no thickness,
and so on. Euclid formulated a geometry that withstood the test
of time beyond 2000 years.

There is a wonderful book by B. K. Ridley called *Time, Space and Things*.[3] I cannot hope to better his words on the subject of 'things'. Here are some quotations from a chapter of that title:

> The living cell is clearly an impossibly complex system, and so, for example, is a surface – any surface. There may be the occasional flirtations with these topics, the one in biophysics, the other in chemical physics, but by and large they are terribly difficult to deal with. Cells and surfaces are not simple things.
>
> It could be argued that simple things plainly do not exist. The Queen might boast to Alice that she could think of as many as six impossible things before breakfast, but she would be hard put to it to think of six *simple* things.

Ridley went on to show that solids made simple by assuming all their faces to be planes could play no part in physics just *because* they suffered from corners and edges: 'Their regularity is only relative. They do not look the same from whatever point of view one cares to adopt. Some directions are "more equal" than others.'

> Getting rid of corners and edges leaves us with *the billiard ball*. To a physicist a billiard ball is a lovesome thing, God wot! It has an archetypal significance in the subject, unrivalled even by the weightless string. It looks the same from all directions and it can be handled, thrown, swung or rolled to investigate all the laws of mechanics. Our stone-age natural philosopher would have insisted on a grant especially ear-marked for the production of spherical stones.* But in spite of the undoubted glamour of our billiard ball, it is still not simple enough. Its flaw is that it has a surface and, as we mentioned at the beginning, a surface is not a simple thing. Yet any object, if it is to be distinguished from its surroundings, must necessarily possess a surface. That being so, we idealize the surface away by pure imagination – infinitely sharp, perfectly smooth, absolutely featureless. And while we are

---

* Ridley had remarked earlier that: 'A stone-age war department might have been keen to commission some customer-oriented research on stone ballistics'.

idealizing, let us make the billiard ball absolutely uniform, of infinitely hard and perfectly elastic material – shall we call it *utopium*? Now there is the first simple thing of physics – a billiard ball made of utopium.

Yet nothing like it exists. The utopium ball is a product of, literally, an ideology, and nobody, being dispassionate, believes entirely in the products of an ideology. That is, paradoxically, its strength. We know from the outset it is wrong.

I have never seen idealised physics so effectively demolished in any language except in Murray Laver's splendid lecture quoted in the next section. And yet, in 13 words he makes it rise like a phoenix out of its own ashes. Later on, he states a truth that should never be forgotten: 'Physics is above all a model-making activity.' As for biology, his warning sentence reads: 'Although an amoeba with a strong personality has yet to be discovered, biology has always to keep a wary eye open for the effect of individual living things . . . trusting that the psychology of the electron as a serious study is a long way off.'*

Yet one need go nowhere nearly as 'far out' as quantum mechanics to arrive at a nonsense produced as a result of blind belief in logic, only, in fact, as far as secondary school geometry. Take an equilateral triangle, as shown in Fig. 2.1. Trisect each side at points A_1A_2, B_1B_2, C_1C_2. On the middle third of each side erect an equilateral triangle outwards, as in Fig. 2.2, erasing the bases of the triangles. The result is the traditional 'star-shape'. Repeat the exercise on each of the straight lines which form the boundaries and you have a shape that begins to resemble that of a natural crystal (Fig. 2.3). Repetition of the exercise a further three times results in the shapes shown in Fig. 2.4(a), (b) and (c).†
However many more stages you attempt, either with a fine pencil, *or in the mind*, the area remains finite and ultimately becomes $1^3/_5$ times that of the original triangle. But each time a new set of

---

* But quantum mechanics is at least a first step towards it! This is one of the most famous footnotes of all time. Edward Teller in *Our Nuclear Future*[4] wrote: 'It has not been necessary to ascribe an internal structure to the electron.' His footnote to this sentence was simply 'Yet'.

† If the ultimate boundary be regarded as a curve, it is known as the 'snow-flake curve'.

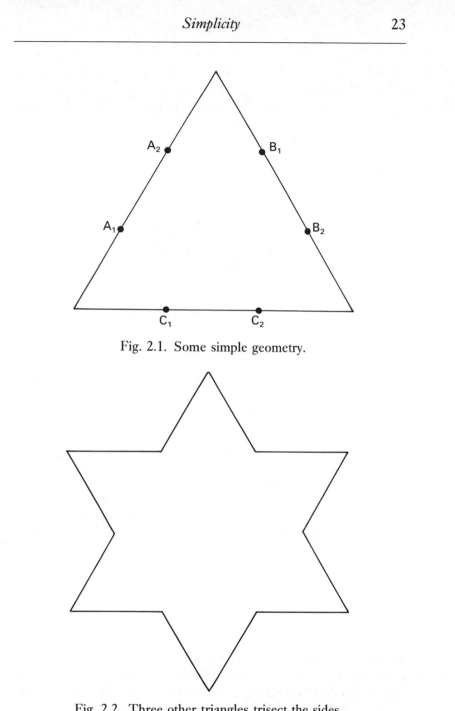

Fig. 2.1. Some simple geometry.

Fig. 2.2. Three other triangles trisect the sides.

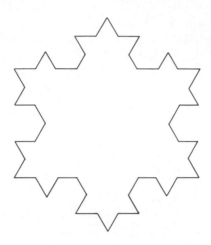

Fig. 2.3. A further stage of development.

trisections is carried out, the length of the perimeter is increased. It is clear that the further one goes, the nearer the length of the perimeter goes towards infinity. What, for a finite area? – and in a planar, two-dimensional figure? Indeed, and what is more, however much the angular shape takes on the appearance of a curve, you will never be able to draw a tangent to it. Its direction at any point is indeterminate. Where now your logic?

What all this adds up to is simple: if you ask a silly question, you can expect a silly answer! All attempts at simplification are steps away from fact.

The Great Bog of 'Un'

During the 1980s, the company IBM ran a series of annual conferences at which eminent speakers addressed an invited audience on an aspect of science and technology in which one particular aspect was emphasised. In 1981 the theme was 'Science and the Unexpected'. One of the chosen speakers was Murray Laver, at

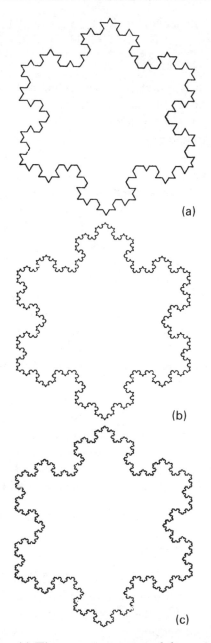

(a)

(b)

(c)

Fig. 2.4. (a) to (c) Three more stages, and the periphery begins to look like a smooth curve.

that time Vice-Chancellor of the University of Exeter, who took
as his title 'The Great Bog of Un'. He explained what it was in
these words:

> I see a small mesa – indeed, it may be no more than a butte
> – rising steeply out of a dreary morass. Its flat top is the
> plane of reason, which includes the field of scientific
> knowledge. Around its base on all sides, to the horizon and
> beyond, lies the Great Bog of Un – the vast quagmire of
> human ignorance . . . I can describe its contents only by
> using UN-words: unexpected, unpredictable, unsystematic
> and so on.
>
> When confronted by the Bog, scientists exhibit a wary
> ambivalence. The scientists I come across rather dislike being
> startled by the unexpected, and accept nothing that can't be
> slotted into their systems – I don't say they ever *force* the facts,
> but they sometimes reach for a heavier hammer.
>
> We created the unexpected by inventing the expected. Sixty
> thousand years ago almost everything must have been
> unexpected. The Neolithic Revolution is the archaeologists
> name for our expulsion from the Garden of Eden. That fall
> from wide-eyed innocence happened when the hunter–
> gatherer's excitement at an unexpected find and glorious
> gorge gave way to the farmer's premeditated planting of crops,
> to the planned and love-less conjugations of his beasts and to
> his discovery of the gloomy pleasures of grousing about
> unreasonable weather, ruinous barter rates and the daft
> decisions of Extremely Elderly Councillors – EEC for short.

Murray Laver concluded with these words of wisdom, a copy
of which I carry with me always so that I can re-read it to lift me
out of moments of despair:

> I have a feeling that Old Mother Nature is chuckling and
> mumbling to herself somewhere out there in the Great Bog as
> she prepares to lob up the next surprise from her inexhaustible
> supply of the unexpected.
>
> I see science as a frail platform built of items salvaged from
> the Bog. We are forever trying to enlarge it by adding new

hypotheses to its edges; though we know that these never last very long. Once superseded they are overlaid by new ones, and the platform rises like one of those mounds of rubbish that so excite archaeologists. Eventually the heap becomes unstable, shifts like a paradigm, and collapses causing great distress to elderly scientists who have given it the best years of their lives.

We raise our young to believe in the myth that new knowledge is never added to the main body until its supporting experiments have been confirmed by others. Everyone knows that, like hurricanes in Hertfordshire, that hardly ever happens: Nobel prizes are not won, grants are not awarded, employers are not amused by plodding repetitions of what has already been announced. There's too much of it about, and it can't be patented. Moreover, although no one believes a scientific hypothesis except its author, everyone believes an experiment except those who performed it – they *know* how precariously and infrequently their apparatus peaked towards the verge of operation.

Haldane once wrote that the universe may not only be queerer than we think, but queerer than we *can* think. Certainly, we are what we are by what our evolution has selected in an environment that clamped our ancestors' attention firmly on food, and on not becoming food; deep thinkers of abstract thoughts were eliminated. It is, indeed, unexpected that so narrowly practical a survival mechanism can be redirected to discover wide and subtle truths, some of which – cosmology for instance – have no obvious reference to our dinners, or to the sex lives we so altruistically endure to keep our species in play.

Evolution for survival has given us senses that respond to changes and are not excited by stasis, so we see 'Life, the Universe and Everything' as one damn transient after another, and seek explanations for changes rather than for steady states. Changes take place in time; and time has always puzzled philosophers – although common sense knows that it is only what stops everything happening at once. Common sense also knew before Einstein did that time is relative; how

long a minute is, depends very much which side of the
bathroom door you are on. Men have always worried about the
start of time, but when I read those racy accounts of inflation
before the Big Bang, I am left with the feeling that there is
something kinky about calling in the biggest and oldest event
we have been able to conceive to explain the smallest entities
we have just thought of – those ephemeral building blocks
now breeding like rabbits, which aren't found in natural
Nature, and which are inevitably doomed to obsolescence,
like all scientific hypotheses.

   When pressed about these incredible matters, scientists fall
back on the defence that theoretical physics must be true or
we wouldn't have all these splendid jet engines and computers
would we? That simply won't wash; it does less than justice
to Rolls Royce and IBM, whose splendid engineers are
practical persons who have always worked on the basis of 'Start
her up, Harry; and see why she don't go'. Any theory was
dreamed up by others, very much later.

Wisdom and civilisation

There is a great gulf dividing Knowledge from Wisdom. Knowledge is a collection of facts stored by the memory and many of
the facts appear to be related by Laws, provided we trim the
'rough edges' off the reality and *simplify* our concepts of real
objects. Any or all of these laws may at any time be swept away
by the discovery of new facts, and we may go exploring in the
hope of doing just this. Experience, curiosity and memory are all
ingredients of knowledge. It is possible that science contains the
whole of knowledge (not that we can ever have access to it), but
there is nothing in the acquisition of knowledge that demands
that we have one grain of wisdom. Those who seek wisdom do
so by the collection of other people's actions, material assembled
by reading the history of the actions of our own kind and by
studying the great literature of the world. Wisdom in fact, dwells
among the Arts (in the modern meaning of that word). The Bible

tells us (Proverbs, Ch. 4, v. 2) 'Wisdom is the principal thing; therefore get wisdom. And with all thy getting get understanding.' The things that can be understood are the reasons for other people's behaviour, in the main. But we shall never understand why masses are attracted to each other. Calling it 'gravity' only names it. Likewise, we shall never understand electromagnetism, not even electric current, in the sense that we understand why actions are taken as the result of jealousy, compassion, greed or almost any of the so-called *abstract* things. Even wisdom was capable of getting its feet knotted, and the dichotomy that resulted in the Monks, on the one hand, and the Spartans, on the other, came about as the result of the Thinker turning its attention to itself and to its body. The one approach regarded thought as 'divine' and the flesh as a hindrance to thought and as a ball and chain to those who sought to move nearer to God. The body had to be subjugated, even disgraced, as a *means* of enhancing the thought process. One wonders just how much wisdom there is in this. The other approach glorified the body as God's ultimate creation. The results of these processes remain with us as relics of a past civilisation – the Monasteries and the Gymnasia. They also remain with us today in much more tangible form, as the people who adhere to either of the two extreme doctrines, much as did our ancestors thousands of years ago.

Wisdom is connected with civilisation, but *that* subject can easily be over-played, as it was in Sir Kenneth Clark's famous TV series of that name. Being himself an artist, he measured civilisation as if it were a physical quantity, and then only in terms of painting and sculpture. Certainly he pinpointed what, for him, were the milestones in civilisation. And in choosing Michelangelo as one of them, he was still only voicing his own opinion. Jacob Bronowski in his *Ascent of Man*[5] saw a very different set of milestones without which the 'arts' would not have prospered at all.

I have heard the milestones in the rise to power of human beings nowhere more persuasively defined and listed than has been done by John Lenihan. In conversation among four people on a train (of which I had the honour to be one), he began what might only be described as a 'sermon' (in the best sense of the word) which began: 'In the beginning there was Technology.'

Humans constructed sailing boats many thousands of years ago, he argued. And the people (for there was no one person working in isolation) who shaped the blade of the plough lifted humankind above a standard of living where every person toiled all his or her life to wrest a living out of the soil, until there was enough food to allow a few members of a privileged class to indulge in the luxury which we now call 'Science'.

It was powerful argument. We are not a superior species because we can make weapons to kill, but because we are permitted the *luxury* of science. It is wisdom that we lack when we use it to wage war. John Lenihan pointed out that he had written a paper on progress from the plough to the horse collar, the horseshoe and the stirrup.[6] It is best reproduced in his own words.

> Two of the most significant and spectacular advances in the history of civilisation were accomplished by technology alone. The military, political and social revolution which culminated (or started) at the Battle of Hastings was precipitated by the introduction of the stirrup. Until the eighth century, a mounted soldier had the advantage of a better view but had to rely on his own muscle power and was helpless when unhorsed. Once provided with stirrups, however, his fighting power was greatly increased.
>
> The horseman now merely held the lance between his arm and his body, allowing the muscle power of the horse to deliver the blow. The feudal system developed as a means of sharing the burden of providing horses and suitably equipped soldiers; the aristocracy enjoyed their lands in return for the obligation of serving the monarch, especially by providing fighting strength. The Franks absorbed and exploited the new technology of war. The Anglo-Saxons did not take it seriously and in 1066 they paid the price of their neglect.
>
> The agricultural revolution which occurred at about the same time was also linked with the technology of horse power. In earlier centuries, the ox was commonly used for ploughing and other agricultural tasks. Horses were not very serviceable, partly because their hooves were easily damaged and partly because their tractive effort was very limited. The

ox yoke was inefficient when applied to a horse because, as soon as the animal took the strain, the neck strap pressed on the windpipe with discouraging results.

The nailed horseshoe appeared during the ninth century and the modern type of harness, with a padded collar allowing the animal to exert full effort, was developed almost simultaneously.

This last paragraph ends:

The rise of the middle classes, the growth of commerce, industry and education and many other features of modern civilisation may be traced back to the agricultural revolutions – technological but not scientific – of nine or ten centuries ago.

One might say of civilisation, as a means of bridging the gap between Clark and Bronowski: if art be the thermometer, then technology is the source of the power. One of the messages I have tried to write in very large letters in this book might be summarised in measuring the experience of science, technology and evolution, thus: science – 200 years, technology – 10 000 years, living organisms – 800 000 000 years.

There must be a great deal more Knowledge, more experience and more wisdom to be found in a study of Life than there is in either science *or* technology.

In an interesting book by L. Sprague de Camp under the title *Ancient Engineers*,[7] one finds a comparison of the different kinds of column that were used to support some mighty buildings of ancient civilisations. Figure 2.5 reproduces the figure illustrating it. The author points out that much earlier in history the columns had consisted of tree trunks with their roots and top branches lopped off, and it seemed 'natural' to mount them as they had been in life, tapering upwards. But some very 'determined' trees had grown new roots, after which the whole thing grew and wrecked the building, so they began the practice of mounting the trunks upside down. The author reminds his readers forcefully: 'Just remember, the next time you pass a bank with conventional Greek columns before it, that you are beholding an imitation in concrete of an

Greek   Greek   Greek
Doric   Ionic   Corinthian

Fig. 2.5. Classical Greek columns.

imitation in stone of a simple wooden log.' Perhaps that is why the ancients succeeded so well in their technology. They copied nature in this and doubtless in many other ways. There are still no equations to describe the shape of a plough blade, just as there are none to define the shape of a blade of grass. Yet both could be said to be more *profitable* than the whole of radio astronomy! One might re-cast a Biblical quotation in almost diametrically opposite terms: 'For what shall it profit a man, if he shall gain the whole world, and lose his own soul?' (St Mark, Ch. 8, v. 36), might become: 'For what shall it profit humans, if they know the Origin of the Universe but lose their ability to use it?'

Our search for ourselves

We have been self-conscious at least as long as we have practised technology. Genesis tells us the story of the Creation and of the fateful fruit that the first couple ate (in the hope of knowing as much as God, is implied). The fact that it marked the beginning of their self-consciousness and therefore, it might be argued, the beginning of psychology, *and also*, the beginning of technology (they sewed fig leaves together to make 'aprons'), I take as mere coincidence, for there are many among us who regard Genesis as a mere fairy story to satisfy the ignorant.

By the time the ancient philosophers arrived, we were well

entrenched 'in the middle' of the universe. In technology, we had to wait until well into the nineteenth century for the invention of the water closet. Both our technology and our philosophy may well be subjects of ridicule by our own grandchildren.

Modern psychologists regard the human brain as being, in part, like the brain of an animal, but also having a facility which might be called 'the advanced brain'. The two facets are said to be almost diametrically opposed in that the animal part is preoccupied with *survival*. It welcomes routine, for routine is security. The advanced brain, on the other hand, demands excitement, adventure and leads us towards taking unnecessary risks. The latter are most obvious in our leisure-time activities, car racing, skin-diving, mountaineering, pot-holing and many others.

The reality is almost certainly not as clear-cut as the psychologists would suggest, but their ideas will serve as a general guide. For example, animals, especially the domesticated ones, appear to have a genuine curiosity, but it is a mere shadow of the human curiosity, the driving force behind our thirst for knowledge, our explorations to the further parts of our lands and oceans, the poles of the earth, the top of Everest and, more recently, beyond the sun to the stars. At the same time we have been almost obsessed with the origin of things, and this curiosity is devoid of the risk element entirely. We want to know how we ourselves came into being in the hope of reassurance about our continued existence after death. There have been gods since there has been civilisation. If there is a brain with imagination, there must be a god to go with it, or one has the feeling that 'the very stones would cry out' (for one).

Yet the churches are not packed to the doors (even on Sundays). Only the imminent threat of nuclear war could do that. Our search for ourselves did not end when we had 'created God in our own image'. (No, this is not a misprint; it is the way it was.) But at least we are turning more towards things of the mind and are perhaps more concerned with other human beings than at any time in history. This is a strange facet of ourselves, for it comes at a time when, superficially, we strive more than ever for material possessions. We are a strange mixture indeed!

3

Human laws and the rules of God

Science again

For centuries the 'pure' scientists threw sand in each other's eyes (and in their own) by pretending that they were looking for 'the Truth'. Some of the pitfalls that they encountered on their journey have already been discussed in Chapter 2. The early days were characterised by a need to divide all things into two opposing categories. Even in theology, there had to be *two* creations. Wherever there was good, there had to be evil, or we would never appreciate good. There was night and there was day. There were pleasure and pain, the quick and the dead, beginning and end. There was black and there was white. There was to be no compromise.

Human laws were essential to community living, if only to demonstrate to our own satisfaction that we were superior to the animal. It was against a background of such laws that science emerged and grew. Much of it was seen to prosper; it became a way of life, one might even say it became a kind of religion, for it had, and still has, many facets in common with the great religions of the world.

If we have seen a good 'whodunnit' play or a film with an excellent plot, it is almost as pleasurable to see it again as it was to see it the first time. At the second viewing one sees it 'in the bright light of hindsight',* knowing who did it and appreciating all the more the ingenuity of the author, an aspect that tended to be forgotten at the first viewing. So it is with science.

'Another parable spake he unto them . . .' No, not about the Kingdom of Heaven, but about science. Science is like English cricket to an American baseball player watching it for the first time. He is fascinated to the extent of wanting to know all about the game, but he is denied any access to the game, except by watching and hearing only such things as can be seen and heard by a spectator who is isolated and prevented from any form of communication with fellow spectators and denied any printed matter about the game. He is simply trying to discover the rules.

Let me say at the start that had he been a tribesman from New Guinea and been introduced to the game by seeing any of the first-class bowlers that an Australian would call a 'chucker', he might be forgiven for concluding that the batsman was on some kind of 'trial by ordeal', and if he made most of his runs behind the wicket, he ran because the fielders threw the ball *at* him!

The American student of the science of cricket would have to watch several matches before discovering the facts behind the roared appeals for 'caught behind the wicket', some appeals sending a player back to the pavilion, others producing nothing more dramatic than the odd oath by the bowler concerned. 'L.b.w.' would take our student a little longer. In each case 'probability' rather than rigid rule would seem to be involved. 'Appeal against the light' would escape him for several reasons, indeed the phenomenon might only occur a few times a year.

Just when he reckoned he knew 'nearly all of it', the following spectacle occurs. Towards the end of a dreary three-day match, two batsmen are well 'dug in', no longer interested in a 'quick single', the field fairly spread. A batsman prods forward and plays a ball along the pitch a matter of two or three yards. No-one can really be bothered to recover it. The kindly batsman walks out of his crease and picks up the ball, but instead of throwing it to the bowler (the latter having turned to walk back) he throws it to

---

* A favourite expression of the professor of my student and lecturer days, the late Sir Frederic Williams, was that in such light, 'All things are obvious'.

the wicket keeper, but with a batting glove on his hand, his aim is poor and the ball hits the wicket. Someone appeals and he is given out. On his way to the pavilion the fielding captain runs to him: 'My dear chap, we want to get you out *properly*, please continue.' The captain indicates to the umpire that this will happen, and the batsman returns to the crease. The American observer shoots himself! (Had the batsman been out, *how* was he out? The possibilities would appear to be run out, stumped, hit wicket or handled ball – answer below).* Cricket, after all, still has a chivalry. Suppose the natural order of things had one, too!

Even scientists today will disagree with the statement that there is only *one* science. The artificially categorised divisions such as physics, astronomy, chemistry and biology overlap to a degree that makes them indistinguishable to layperson and expert alike. More than this, we have long realised that what at first appeared a dichotomy could be only two facets of one and the same phenomenon, and there are probably more facets yet to be found. The early students of 'light' asserted that it consisted of streams of particles – 'corpuscles' they were called. Properties for these (later to be asserted 'non-existent') particles were invented by the human mind in terms of the properties of other things that it knew more about and were found satisfactory to explain all known phenomena exhibited by light – until someone pointed out the colours produced by an oil film on water. Science waited for Thomas Young to unravel the mystery with his wave theory, and when *that* was found to satisfy all known events, including interference and diffraction, the corpuscular theory was thrown out of the window with the proverbial bath-water, and science waited even longer for Max Planck, Niels Bohr and the Braggs and their demonstrations that light came in 'packets of waves' (whatever they were) and that electrons could be diffracted as if they were a pure radiation. But the message was indisputable. Sometimes one needs the corpuscular theory, sometimes the wave. They are both facets of a greater whole.

---

* Actually he is out 'bowled'. Consider what the ball did. It hit his bat, then the ground, then his hand (which counts as the bat), then the wicket. This is indistinguishable from 'played on', i.e. bowled.

I have never seen the subject of waves and particles contained so succinctly as in the following words of Sir Lawrence Bragg:*

> Electrons scattered by matter exhibit interference patterns of a form which would be accounted for by their being waves. Particles behave like waves, and waves like particles.
>
> The relation is a subtle one. When we postulate an experimental set-up, and wish to prophesy the result, we must treat both light and matter as waves. The nature of physical reality is such that we can only calculate the relative probability of an effect in various places. On the other hand, when we write the history of what did actually happen in an experiment, it is a history of particles whether of matter or light. A wave-like uncertain future, only expressible in probabilities, forever streaming through the moment 'Now', is transformed into a definite past of particles. Determinism takes on a new meaning.

In the even *brighter* light of hindsight we can begin to see some of the most amazing facets of the study of the universe through the eyes of a scientist. The 'new look' probably can be dated at Heisenberg's discovery of the 'Uncertainty Principle'. If you know where a particular fundamental 'bit' of matter is located at a given instant, you *cannot* know its velocity, whereas if you know its velocity you cannot know where it is. The future was as uncertain as ever.

Perhaps even more amazing than the Uncertainty Principle itself is that it can 'live' alongside the knowledge that for 99% of our daily lives, the laws of the universe suggest that there will always be the same unique result from the setting up of the same set of initial conditions. Objects lifted above the surface of the earth and released have always fallen to the ground at roughly the same initial acceleration. Should a child ask for an explanation of this phenomenon, the 'learned' of the 1990s are still likely to reply, 'It is subject to the law of gravity.' Let us be quite clear

---

* From notes produced for a science course for civil servants held at the Royal Institution, London in 1968. For the preceding paragraph of this quotation, see Chapter 6, pp. 177 and 178.

about this one fact. *There is no proof that it will happen next time!*

So there is *no* law of gravity, *no* Ohm's law, or whatever. They are all collections of known phenomena that have been classified and labelled in exactly the same way that a botanist might collect herbs. The only difference is that when taken together in mixtures and combinations, the laws of physics almost always appear capable of predicting the result of a given set-up. But be ever aware, it *is* only 'almost'! The principle of 'the same answer to the same question' has been called delightfully by Herbert Dingle: 'the Uniformity of Nature',[1] and earlier, by Sir William Bragg, 'the Nature of Things'.

As far back as 1883, Professor Osborne Reynolds included this warning in an address to the Society of Arts (now the Royal Society of Arts).

> Science teaches us the results that will follow from a known condition of things; but there is always the unknown condition, the future effect of which no science can predict.

Science brought with it a rigid way of doing things. 'The scientific method' more than suggested that there was only *one* way of going about any exploration (and what is life if it is not an endless series of explorations?). But at least *one* successful scientist put the record straight concerning the method of procedure. In his delightful book, *Induction and Intuition in Scientific Thought*, Sir Peter Medawar wrote:[2]

> Ask a scientist what he conceives the scientific method to be, and he will adopt an expression that is at once solemn and shifty-eyed; solemn, because he feels he ought to declare an opinion; shifty-eyed, because he is wondering how to conceal the fact that he has no opinion to declare.

Martin Gardner quotes 'an unknown physicist' writing in *The Times* to the effect that 'nuclear physics had been battering for years at a closed door only to discover suddenly that it wasn't a door at all – just a picture of a door painted on a wall.'

Things of the mind

Having, as it were, undermined the laws of physics, at least in the context in which they have always been introduced in school physics teaching, one cannot simply walk away and pretend to be unmoved by what one has been taught. There *are* patterns in nature, and by 'nature' I mean every phenomenon in the known universe, animate or inanimate. It is just that in the 1930s, school physics was divided into subjects and these were studied by the use of quite separate books. There was a book on 'heat', another on 'light', a third on 'sound'. They managed to combine 'magnetism and electricity' in a single volume. Finally there were 'elementary mechanics' and 'properties of matter'. This last book was devoted to all the oddments that didn't seem to fit into any of the other books, topics such as viscosity, surface tension and osmosis. All I have done in my apparently vicious attack on physics is to say that *all* the books could be combined under that one last title 'Properties of Matter'.

One of the most basic ideas is that of liquids that flow, and of what causes them to flow. As soon as measurements are made on a fluid circuit – and measurement is the very bedrock of all good physics – the investigator will soon discover that *pressure* causes *flow*. Pressure is simply another way of expressing driving force: in a water pipe it is, in fact, the force per unit area of pipe internal cross-section. The thing that restricts the flow is the roughness of the bore of the pipe coupled with the viscosity of the liquid which is flowing. Experiment will soon reveal that the resistance to flow increases in proportion to the length of the pipe and inversely as its cross-sectional area.

Now, *force* is a thing we can experience physically – we can feel it. The flow of water we can *measure*, so that most of the quantities are, so to speak, real and tangible. But when we deal with electric current we are on much shakier ground. Historically, electric current theory began with mixtures of chemicals which were seen to cause things to happen to magnets placed near the wires that dipped into the chemicals. They guessed it was due to a flow of 'something' in the wires and Ohm's law was born:

$E = iR$, where E was a kind of electrical force (electromotive force, e.m.f., in fact), i was the flow rate of whatever it was, and R was the 'resistance'. This last quantity varied from material to material, but was nevertheless shown by experiment to increase with wire length and inversely as the area of cross section of the wire, just like the resistance of water flow in the pipe. What *is* disturbing to many is that neither e.m.f. nor current is real and tangible, but the same formula can be used to good effect to predict the flow due to a given e.m.f. in a given wire circuit, *as if* the flow existed.

One is reminded of Tyndall's statement in respect of the principle of gravitation: 'If Dalton's theory, then, accounts for the definite proportions observed in the combinations of chemistry, its justification rests on the same basis as that of the principle of gravitation. All that can in strictness be said in either case is that the facts occur *as if* the principle existed.'*

Here we move into engineering, one might almost say, 'as *opposed* to physics', for engineers are concerned with setting up a collection of rules that *work* (for a given range of situations). They are concerned only with whether the concepts (things of the mind) remain valid to predict what will happen when certain real situations are set up, so that by processes of trial and error, experience and intuition, they can profit from the concepts by arranging for desirable things to happen. They have become *designers*.

After 40 years of research into the phenomenon we call 'electromagnetism' I am well aware that I do not know how to teach my subject. All I can do is to take specific examples of apparatus and, by demonstration, hope to enlighten my students about the subject until one day, hopefully, the 'penny drops', for each of them, as it did for me (even though it be long after they graduate, as it was in my own case). After that they will find a new excitement in electromagnetism, realise its still untapped potential and yet know with certainty that such things as magnetic lines of force, e.m.f.s and even current itself, are all figments of the mind.

Let me give you an example, which, being essentially mechanical only in nature, will hopefully bridge the gap between the

* John Tyndall commenting on Faraday's Royal Institution Discourse.[3]

tangible and the intangible. Figure 3.1 shows a row of steel rods, some bent into a right-angled shape, each of which is free only to slide up and down in a pair of holes. The rods can be made to rise and fall by means of cams operating on their lower ends. All the cams are locked together on a common shaft in such a way that when the shaft is rotated, the tops of the rods give an impression of a travelling wave, not unlike waves in the sea (before they break on the beach) or ripples in water. But what does *actually* travel along as we turn the handle is *nothing – no, real, thing*! By the very design of the apparatus we have made horizontal movement impossible. What gives the *illusion* of movement is that we have seen water waves on many previous occasions and have assumed that water *did* travel shorewards. So we mentally join the rod tops with an imaginary line, and it is this line *only* that moves. But we have missed the vital bit. If a table-tennis ball is dropped into the trough of one of the waves whilst the system is moving, a *real* horizontal ball movement will occur. The transition from a concept of the mind to a physical reality, as we know it, is amazingly simple if you can only find a good example, such as this, for the action demonstrates that of an electric induction motor whose rotating part need not be physically or electrically connected to anything in order to experience a driving force. I know of no more convincing way to 'explain' an induction motor than with the aid of this apparatus.

But it is not only in electrical phenomena that things are not always as they seem. Figure 3.2 shows a pair of gearwheels in mesh. The larger wheel has exactly four times the diameter of the pinion and four times as many teeth. If each wheel is pivoted at its centre, a single rotation of the big wheel will cause the pinion to rotate *four* times in the opposite direction. But if the big wheel is held fixed, and the small wheel is rolled around it so that the teeth are always in mesh, how many revolutions will the small wheel have made *on its own axis* by the time it has completed one orbit and arrived back at the starting point? (If you wish to try and work it out, do not yet read the footnote on p. 46.) Let it suffice to say here that it is certainly not four!

If you were to meet this same phenomenon in relation to

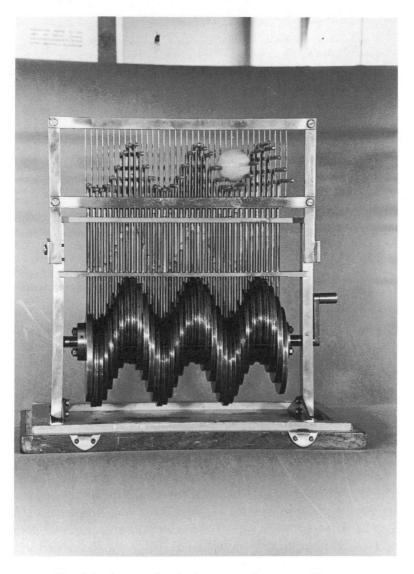

Fig. 3.1. A row of rods demonstrating a travelling wave.

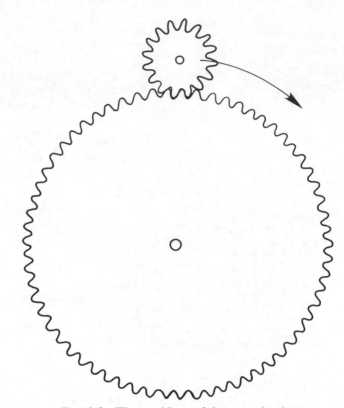

Fig. 3.2. The problem of the gearwheels.

angular *velocity*, as opposed to merely angular position, as is really the case with the gears, you might find the problem considerably more baffling, but there is an experiment that can readily be set up. All you need is a chair seat suspended from a stout tree branch or a beam or a hook in a ceiling by means of a single rope or wire. Sit quite still in the seat until all rotations due to stretch and twist in the rope have ceased. (An assistant can help you attain this situation fairly rapidly.) Now rotate your fist in a horizontal circle. Your whole body will at once rotate in the opposite direction, continuously, so long as you keep the fist rotating. The instant the fist stops, your whole body stops dead. Now astronauts know this. They can re-orient themselves in space just by merely moving a hand or a foot in a circle in any desired plane. But it is a weird effect, to say the least, for the circle made by the hand or foot never encloses the mass centre of the body or the axis about which it rotates!

Yet these are only two-dimensional problems. In three dimensions the complexity is several orders of magnitude more difficult. Consider a helter-skelter. Imagine that you are driving a car up the helter-skelter and that you have an ideal 'spirit level' (which is not affected by centrifugal force, unlike a *real* spirit level)* fixed athwartships of the car instrument panel. At all times the instrument will assure you that the car is laterally level. Yet in fact the 'road' is continually rolling outwards (viewed fore and aft) as you ascend. Proof? Make a coil of spring wire as shown in Fig. 3.3. Paint a mark on it at X. Holding one end of the spring Y securely in a vice, pull the other end Y' so as to stretch the spring. The point X will rotate in relation *both* to the axis of the spring and to the axis of the wire itself. Is not the stretching

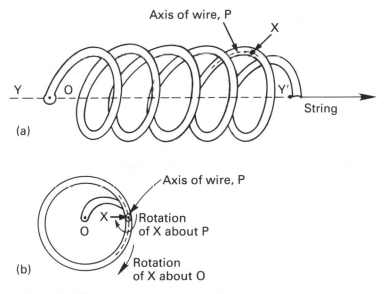

Fig. 3.3. What happens when a coiled spring is stretched. (a) Side view. (b) End-on view.

* A gimballed gyroscope will do this perfectly, if constrained to be 'earth bound'.

of a flat strip spring the only ingredient needed in making the helter-skelter?*

But a more convincing proof is to be found in the principle of *reductio ad absurdum*, used to great effect by engineers in tight corners. What it involves is simply the process of modifying the problem by taking it to ridiculous extremes. Imagine a helter-skelter in which the pitch of the helix has been stretched to many times the diameter of the central column. Now it is easy to see that rotation around the central column has become almost all rotation about the fore and aft axis of the car.

When one starts to enquire into what is truly 'fundamental', one may arrive at a quite different answer from that in the standard physics book. 'Mass, length and time,' says the latter. 'Force, length and time,' said Faraday and Maxwell. I agree with *them*. But let us probe even deeper and ask whether these most basic things are themselves related. In particular, let us examine length and time. Let us try to explain 'length' to an alien being who may not see the universe as we do. We may well end up by saying that one can only ascribe a meaning to *relative* lengths (i.e. how many times one length will be contained in another), and if pressed for an absolute reference length we could perhaps define it as the length we could run in a given time, i.e. we have 'personalised' it. After that we can define time in terms of how long we

---

* Solution to the gear wheels problem on p. 42. The small wheel makes five revolutions in whichever direction you choose to roll it. It makes four because of its teeth ratio to the big wheel and a further one because it has bodily revolved around the centre of the big wheel. Unconvincing? Of course. But look at Fig. 3.4. We have marked the point of contact at the start as A and B on the two wheels. In (a) the pinion has made one complete revolution because the point A is nearest the bottom edge of the page. And obviously it has some distance to go before A makes contact with the wheel again. How much distance? Figure (b) shows that A will contact the big wheel when one quarter of the teeth of the big wheel have been traversed as shown. But in relation to the bottom of the page (and from now on we shall refer to the bottom of the page as ourselves – the 'external observer'), Fig. (b) shows us that the pinion has made $1\frac{1}{4}$ revolutions as the external observer sees it. By the time the point again contacts the big wheel, the pinion will have made two revolutions plus two quarters, i.e. $2\frac{1}{2}$ in all (Fig. (c)). At '9 o'clock', as in (d), it has made three revolutions plus three quarters and it finally returns to its starting position having made four revolutions plus four quarters, i.e. five revolutions in all.

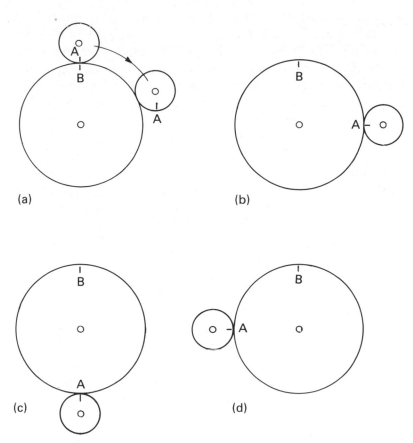

Fig. 3.4. (a) to (d) The solution to the gearwheels problem. See footnote to p. 46.

live, for this could again be personalised in terms of one's own experience to date. Finally, the alien might ask us about numbers and what we mean by 'one'. Having the notion of personalisation in mind, it is now no problem to define number. All that each of us needs to say is: '*I* am *one*.'*

In the mid-1960s I had a three-minute spot in a weekly TV programme for children in the 10–15 age range. I took it as a challenge to show them that such mighty topics as relativity could

---

* i.e. I the thinker.

be made to appear as nothing more than the application of common sense. There are things you can do with TV or film that cannot be done in everyday life. For example, if a performer points directly at the camera lens, he points at *every* viewer, and it makes no difference whether the viewer sits directly in front of the viewing screen or not. The alien might be surprised to learn that two people on opposite wings of a TV set would see the *same* finger pointing directly at each of them. Yet we have not experienced difficulty in adapting to such a situation.

Another possible exploitation of the TV medium which I used on the occasion just mentioned was to deal with the question of relative length, and in the process to illustrate personalisation and the relationship between length and time, all in three minutes!

The scene began with a view of a cube which almost filled the screen. It was resting on ground that was made up only of sand and stones (of indeterminate size, therefore). My voice asked, 'How big is this cube? It could be a very small cube, very close to the camera, or a very big cube in some huge desert.' As I was saying this, the camera pulled back slowly, revealing more and more of the scenery around the cube, and the first recognisable object to appear was a fully grown tree. You could tell it was fully grown because of the gnarled and twisted bark and the complexity of the branches. 'Ah, so it was a very large cube,' I said, 'unless – it was a very small tree.' At this point I put my hand on the cube and all but obscured all of it. The tree was, of course, a Bonsai tree of great age but small stature. 'Unless,' I continued, 'it was a very large hand!' But by now they could see the whole of me and identify the voice with the mouth movement. For purposes of size comparison, *I* was the ultimate 'ruler', i.e. the viewers themselves could identify themselves in me.

Next I pointed out that there was another means by which they could have known that it was a tiny tree. There could have been a wind blowing on it. A close-up of the tree blown by a fan (not seen) at once gave the impression of smallness, without making any 'scientific' deductions. The viewers had experience of fully grown trees in a wind and their branches *sway* under such conditions. Those of the Bonsai tree merely fluttered. So the *timing* helped us measure the *size*. **But,** I then had the tree filmed with a high-speed camera so that I could show its movements in slow motion, and as the film was slowed down, the tree (the only object seen on the screen) simply 'grew' before the eyes of the viewers. I shall have more to say about the time/length relationship in Chapter 4.

The laws of planet earth

Returning to our 'properties of matter', let us look at their relationship to the everyday experiences on earth, as opposed to, possibly, those phenomena that will apparently go with us through the entire universe. For practical reasons, if for nothing else, these latter phenomena must essentially be 'things of the mind'. The laws of planet earth can all be laid down in terms of properties of matter, to which the earth contributes only three of its own. One is its size (related to length) and density. Some would say: 'You mean its *mass*,' to which my reply would be: 'Not really, and I suppose I didn't mean density *either*. What I meant was the force that it exerted on loose pieces of stuff on its surface.' The second property is its rotational speed, both on its own axis and in its orbit around the sun (related to time), and the third is its temperature (related to the sun's properties and to distance). We will examine the laws of planet earth from a number of points of view, for example size. This will also, incidentally, serve as an introduction to Chapter 4, without stealing any of its 'meat'.

There is a maximum size of bird that can fly. There is a logical argument for this which, unlike some we have discussed, does appear to hold good in the real world. It goes like this. When evolution has done its best to develop the finest structures intended for flight, we see the birds. They have hollow bones, for maximum strength : weight ratio, an enormous breast bone and huge shoulder muscles. They have retractable landing gear (legs and feet), tail feathers that can be tilted downwards to provide 'air brakes' (the 'flaps' of aircraft). There are wings specially adapted for carrying large weights, mostly for gliding, using existing air currents for lift. There are wings for hovering (the humming birds, and in larger sizes, the hawks), wings for silent flight (the owls) and even wings that can be used as weapons (the swans). But when the best 'designs' have been achieved the simple limitation of size puts a clamp, even on the apparently infinite ingenuity of evolution.

The lifting force is proportional to wing *area* and therefore to (length)^2.

The weight to be lifted is proportional to *volume* and therefore to (length)^3.

Take a sparrow and double its size in every dimension. It will have four times the lift but eight times the weight. Its wings will need modification. But there comes a limit to modification, and an ostrich will *never* fly, however its wings are modified. This family of birds, which included the mighty Oepyornis that roamed the island of Madagascar up to a mere few hundred years ago* and stood some 15 feet tall, can bring you encouragement or dismay, depending on your outlook.

Alas poor ostrich, your wings were partially shrunk along the road of evolution. But you were not entirely alone. It is interesting in all these possible/impossible situations to examine those creatures that have gone to the very brink of the abyss. In the case of flight, it is a seabird whose plight brought it the nickname 'South Pacific Booby'. The BBC have in their archives a splendid video tape of this bird taking off and landing. Those of us who pilot or have piloted aircraft will recognise all the facets of our artificial control systems.

First the bird needs a runway – a stretch of flat sand at least 50 yards long. Its feet were made for swimming so they serve it very badly as 'wheels'. The wings flap and hit the sand with some violence. The *power* seems inadequate. There is a real chance of failure. Suddenly, perhaps, a short headwind gust. The undercarriage is retracted, the flapping stops, and from then on the

---

* The famous broadcaster David Attenborough reconstructed a complete egg of *Oepyornis* from fragments of shell collected for him by natives of Madagascar. It measures some 34 inches (86 cm) around its long axis and has a volume equal to seven ostrich eggs or 168 hen's eggs (10.68 litres or 2.35 gallons!).

elegance is undeniable. Gliding for 95% of the time, the bird was *obviously* perfection as a flying machine. Unfortunately what goes up must come down and if the take-off was crude and risky, the landing is a near disaster. The wing feathers are put in the 'stall' position. Down come the air brakes (the tail feathers). Air-speed must be reduced, more – more! The undercarriage comes down, the soles of the feet turn forwards to take the impact, but to no avail. The landing speed was too high. The bird rolls head over heels, neck all twisted to one side. Covered in sand, fortunate not to have broken its neck, it picks itself up ruefully. Evolution has met its match. The laws of planet earth boom out: 'Not a tenth of an inch bigger!'

Properties of matter

From the primary properties of planet earth, its temperature, its speed and its force of gravity, we can deduce consequent properties as follows.

The nature of the elements is such that at the surface temperature of 'Mother Earth' some elements are solid, some liquid and some vapour (gas). It is all a question of the degree of molecular or atomic mobility. One rule that I, as engineer, have found unshakable is that in studying the properties of matter on the one hand and the universe as a whole on the other, there is never anything 'black' or anything 'white'. There are only *shades of grey*. Thus there is no clear dividing line between solid and liquid, nor between liquid and gas. There is equally no clear demarcation between plants and animals. The 'borderline' creatures are always fascinating. On the question of solids, liquids and gases, a classic experiment performed decades ago consisted of drilling a small hole near one corner of a large cube of pure lead and pouring into it a small quantity of hot liquid gold which immediately solidified. Years later the diagonally opposite corner of the cube was cut off and tested for gold content, and sure enough, some gold had 'diffused' through the lead as if both metals had been liquids all the time. An equally famous experiment often quoted in school physics texts is one in which a meniscus marking the boundary between liquid in a test tube and gas above it can be made to 'vanish' by choosing the right temperature and pressure, at which point one cannot say whether the whole tube is then full of liquid or of gas.

The lightest gases, hydrogen and helium, have such intense atomic activity (motional energy) that by a succession of accidental collisions with other particles and with each other an individual atom, from time to time, acquires a velocity of over 25 000 miles per hour (40 000 km per hour) away from earth. Now, this velocity is important because it is just sufficient to carry a body completely out of the earth's gravitational field, so long as it meets no other obstacle on the way.* If you liberate hydrogen in a laboratory, it is certain that in 20 years, 99% of it will have left the planet for outer space. The heavier gases, which include nitrogen, oxygen and carbon dioxide, cannot escape from the gravitational pull by random collision. The moon, of course, has insufficient gravitational field to retain *any* gas. So just one fundamental of our planet earth not only fixes the constituents of the atmosphere, it very largely governs its other properties:

> density at ground level,
>
> pressure at various levels,
>
> viscosity at ground level.

It is these 'secondary' properties of matter that have made possible the life cycle of the living things and that have produced the various laws of planet earth, of which only the flight of birds has so far been discussed.

The life cycle on earth is based on the element carbon for its

---

* In the 1950s when arguments raged about the possibilities of humans going into space, some very strange statements were made by otherwise quite well-qualified scientists. Statements that no human could withstand the acceleration needed to attain 'escape velocity'. If we had a fuel with a thousand times the calorific value of modern rocket fuels and time to waste, there is no *physical* reason why someone should not go all the way to the moon at 10 miles an hour! (Surprisingly, it would take only 3 years.) With atoms the story is quite different. They can only rely on a very small number of impulsive collisions within the relatively 'thin' layer of gas we call the 'atmosphere'.

structures, water for its bulk and re-cyclable nitrogen for its food. We seldom stop to marvel at the vital roles of the properties of water. It is a fact that water has a maximum density at a temperature of 4 °C, not 0 °C. Imagine a world in which it were otherwise – one in which the density increased down to 0 °C, or one in which ice was denser than water. As a *container* of energy in the form of heat, water has no equal. Most metals can only store the same amount of heat in about 10 times the same mass. In changing ice into water, 80 times as much heat is required as is needed to raise its temperature 1 °C. To change water into water vapour (to 'evaporate' it, as from the surface of the sea) to make steam requires 240 times the amount of heat needed to raise its temperature 1 °C. To appreciate just how vital are these statistics, one need only read a good relevant piece of science fiction. One that was recommended to me concerned inhabitants of the planet Venus, whose atmosphere was so hot that liquid water never appeared. The Venusians breathed sodium vapour. They made a spaceship and, of course, its temperature was kept at around 500 °C. They visited earth, orbited it and wondered why the earth-dwellers had elected to make such a large area of their land absolutely *flat*. They tried to land on the flat plain many times and each time they touched down they lost so much heat in a few seconds that all they could do was to limp home to Venus. They had never known *water*!

Sometimes we need science fiction to bring home to us many daily occurrences that we tend to take for granted.

Viscosity

We are by no means finished with the subject of flight by discussing the maximum size of birds. At the other end of the scale, there is a minimum size of insect that can fly, i.e. propel itself in still air. Many smaller insects have wings and use them for movement from place to place, but they merely use them as sails. They wait for the right prevailing wind and launch themselves into it. This is a viscosity effect, pure and simple. As creatures are scaled down they see and feel the air as 'thicker'. By the time they are the size of some Microlepidoptera (which include clothes moths), their wings have changed their form from that of a membrane stretched on a frame and covered with scales. The tiny moth's forewing has become nothing more than a stiff bristle carrying a tightly packed row of hairs that stream out behind on the forward stroke but 'feather' (a rowing term borrowed from nature and inappropriate in the sense that I use it here), i.e. splay out, on the return stroke so as to present an oar-shaped blade that resists movement through the fluid we call air, but which the moth would undoubtedly call 'water'. The air appears as viscous to the small insect as water does to us and it literally 'rows' through it, but because of the high frequency of the wings we call it 'flight'. By the time you shrink an insect until its wingspan is less than 2.5 mm (0.1 inch) the 'liquid' (air) has degenerated into treacle!

Gulliver's travels

In my professional capacity of electrical engineer, I wrote a paper in 1965 in which I derived a dimensionless factor, which I called the 'Goodness Factor', that could be used to assess the profitability of making electric motors of various shapes and sizes.[4] It emerged that when electrical machines had electric currents flowing in both moving and stationary parts, their design would automatically become easier and the products 'better' (in whichever coin you chose to take the profit) as they were made bigger. In the same year, Russian engineers developed the same factor and called it the magnetic Reynolds number, to emphasise its importance as the very foundation of the subject.

Now, the Reynolds number, named after its originator Osborne Reynolds, is concerned with the boundary between the two kinds of fluid flow: 'laminar' or smooth flow in layers, as opposed to 'turbulent' or irregular flow. Both Reynolds number and magnetic Reynolds number increase *naturally* with increase in size. In 1973 I wrote another paper on the essential difference between electric motors with electric current in both moving and stationary parts,

and motors with current in only *one* of those parts.[5] The latter I
termed 'magnetic machines', to distinguish them from 'electromagnetic machines' with currents in both members, pointing
out that magnetic machines get better the *smaller* they are. In
other words, the Lilliputians only made magnetic motors,
whilst the dwellers of Brobdingnag only made electromagnetic
machines.

Clearly, without knowledge of these works, a physicist, Dr
Wolfgang Gloede and a geologist, Dr Klaus Wunderlich, wrote,
in 1976, a remarkable paragraph in their book *Nature as
Constructor*:[6]

> In the region of large Reynolds numbers it is the acceleration
> work (inertial forces) that predominates. On the other hand,
> small Reynolds numbers indicate more frictional work has to
> be performed – the viscosity of the medium (the air)
> acquiring more influence here. Since in small insects both
> the wing length (*l*) and the flying velocity (*v*) are small, the
> Reynolds number also becomes small. This means no more
> than the fact that the tiny flyers already begin to feel the
> viscosity of the air to a considerable extent. Under this aspect
> even the bizarre wing forms (such as bristle wings) become
> comprehensible: the prevailing flow conditions are simply
> quite different from those obtaining in the case of their big
> flying cousins or even of birds. In these dimensions profile
> and curvature of the wings lose their sense, for there arises
> hardly any lift at the wing. Insects row in the air just as beetles
> in the water.

Thus it appears that the turbo-alternator of the modern power
station, delivering over 500 million watts with a high Goodness
factor corresponds to the albatross and its kind, operating at a
high Reynolds number in a flight as effortless as that of the
unpowered glider. At the other end of the size range, the moths
that are classified as 'Microlepidoptera' have, like the designers
of electric clock motors, abandoned high Reynolds number and
capitalised on what is virtually an entirely different mechanism
of dealing with fluids, which only works economically in small
sizes.

Flight is not the only ability in which viscosity has the final word as to what sizes of creature 'can' and 'cannot'. There is a maximum size of insect that can live on planet earth. This is set by the insects' method of breathing. Instead of lungs they have a series of long tapering tubes called 'tracheae'. Air cannot flow in such small pipes and therefore the necessary oxygen can only reach the innermost parts by diffusion, i.e. the random movement of jostling atoms. Tracheae 1 micron (1 ÷ 1000 mm, a micrometre) in diameter will serve a muscle of some 7 to 15 times that diameter when the oxygen intake of the muscle is from 1.5 to 3 milligrams per gram of muscle per minute, which is the order of intake achieved when an insect flies. Hence a muscle fibre in which tracheoles (the ultimate branches of the trachaeal system) are restricted to the outside of the fibres cannot exceed 20 microns in diameter. The most powerful dragonflies approach the maximum.[7] The largest insects are the Goliath beetle, *Goliath giganteus*, for volume and weight (100 g; 3.5 oz), and the female Hercules moth from New Guinea for wing area (645 sq. cm, 100 sq. in.). For sheer length, the giant stick insect, *Pharnacia serratipes*, takes the prize at 33 cm (13 inches) long, whilst the largest wingspans are the 30 cm (12 inches) of the Brazilian Owlet moth, *Thysania agrippina*, the dragonfly, *Tetracanthagne plagiata*, and the long-tailed 'Moon moth' from Madagascar, with its 18 cm (7 inches) long tails. Fossil remains tell us that a dragonfly with a wing-span of 70 cm (27.5 inches) once flew over planet earth's surface, and although this is undoubtedly the largest overall insect ever, its body thickness and weight are less than those of the Hercules beetle or of the Goat moth, *Xyleutes boisduvali*, from Australia, which has a body some 12 cm (5 inches) long and over 2.5 cm (1 inch) diameter.

Gravity alone

When it comes to the mammals, viscosity plays no part in limiting the maximum size of lung. It is the primary property of planet earth, gravity, that determines the maximum size of mammal, for the structure of bone, determined largely by the carbon atom, rules that as size increases there comes a point where the weight of a mammal would crush its own bones if it tried to stand on four columns of bone. The prehistoric hairy mammoth was right on the limit, and its size was many times that of the largest elephant of today.

What is also interesting in relation to the size of mammals is that, in general, the larger the mammal, the greater proportion of its time it takes in masticating its food. It has been estimated that the mammoth spent 22 out of 24 hours in such activity, in which case it must have been asleep for some of the time (which is clearly not impossible). In making such generalisations, however, one must distinguish between the carnivores and the vegetarians. Barbaric as it may be, flesh is more readily digestible than plant tissue and there are few vegetarians who manage to get all the goodness out of their food in one go. Rabbits eat a proportion of their own droppings in order to pass it through the system a second time. Cows and many other animals 'chew the cud', i.e. they regurgitate the material swallowed into their mouths and give it a second pounding with the molars. Rabbits, incidentally, also chew the cud as well as eating it a second time, which illustrates just how difficult it is to break down the cell walls of plants. Vegetarians eat frequently throughout a day. Carnivores, on the other hand, tend to eat massive meals spaced well apart. The giant cats will feed perhaps once in three days, and potential victims react accordingly, mingling with their predators when the latter are seen to have just fed.

Having defined the maximum size of mammal on the grounds of gravitational constraint alone, be sure that gravity also plays a contributory role in matters concerning flight, the terminal velocity of objects in free fall (as in Chapter 4, pp. 63–64), the maximum height of trees, the spread of their branches and many other 'top and bottom' limits as imposed by the properties of matter and the dimensions of the earth.

Some of the prehistoric reptiles, such as Brontosaurus, appear to have broken the size barrier in respect of land-going creatures. But it must be remembered that they took a large part of their weight on their bellies to relieve the stress on the bone-filled legs. In the ultimate, the entirely sea-borne mammals (whales) distributed their load uniformly along their entire length and there is no limit on the size of a whale imposed by gravity alone.

Now, there are engineering limits that are every bit as restrictive as each law of planet earth. There is a maximum length of chain that can be hung from a fixed hook, for there comes a length beyond which the total weight of chain exceeds the breaking strength of the top link. This is not of any great consequence in the erection of buildings by cranes, for high-tensile steel with a breaking strength of 170 tons/sq. in. (264 000 tonnes/sq. m) and a density of 8900 kg/m^3 (500 lb/cu. ft) could reach a length of about 13 400 m (44 000 ft) which is higher than Everest and 35 times the height of the Empire State building.* But in water the limit is more tangible, especially in the subject of sinking diving bells and the like into the deepest parts of the ocean (the Marianas Trench off the West Coast of South America), for the ocean is deeper than Everest is high. The weight of an object in water is less than that in air by the weight of liquid displaced (a fact first recorded by Archimedes) but this only relieves the weight of a steel chain by about 13%. Nevertheless, ultimate chain lengths are of the same order of magnitude as the greatest depths in the ocean, and a single steel cable extends the maximum possible suspension by a factor of 2.5, i.e. to about 38 000 m (125 000 ft),

---

* Arthur C. Clarke, in his *Fountains of Paradise* (1989, London: Gollancz), postulates a fibre, made in outer space from pure diamond crystal, that has such tensile strength that an invisibly thin strand of it will support the weight of a man. This is the only 'miracle' he asks the reader to accept in a marvellous science fiction story in which the rest all follows naturally from the known laws of physics. With this material he builds a lift shaft one third of the way to the moon. In a lecture at the Royal Institution in London in 1980, to commemorate the late Dr Chris Evans, Arthur C. Clarke took as his title 'How near are we to the Space Elevator?'. He pointed out that if a suspended fibre is made thicker the nearer it is to the top, *any* material could be used, except that there may not be enough of it on earth! But he went on to say that since the gravitational force reduced with increased height, the task was easier than a superficial examination might suggest and that our technology in 1980 took us 60% of the way.

Today's science fiction is tomorrow's science fact.

but it is clear that other substances would not meet the requirement, mild steel, for example breaking at just over 5700 m
(19 000 ft) as a single cable* (see Fig. 3.5).

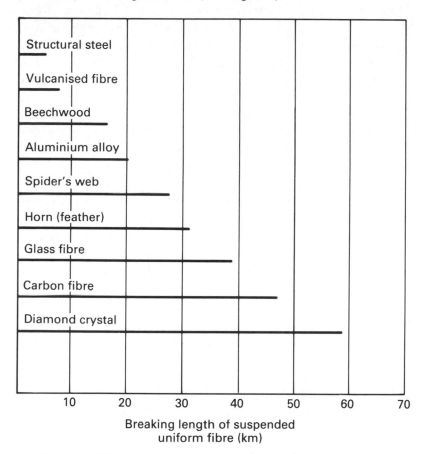

Breaking length of suspended
uniform fibre (km)

Fig. 3.5. The maximum lengths of strands of different substances that can be hung without breaking under their own
weight.

* The tensile strength of a spider's web is, based on this criterion, over
five and a half times the strength of steel, for it could suspend over 27 500 m
(90 000 ft) of itself, and is only bettered by nature's horny material (as in
feathers) or by glass and carbon fibre and diamond crystal (see Fig. 3.5).

Earth rotation

If one of the rules of planet earth is to set a minimum size at which a creature can fly, it is reasonable to ask why no bird has come within a factor of 10 of such minimum even in *linear* dimension. For that matter, so also could one ask why the smallest mammal is thousands (perhaps tens of thousands) of times the weight of the smallest insect, which insect also walks on the ground, albeit on six legs rather than four.

It is something of a surprise to find that it is the speed of rotation of the earth on its own axis that has set these size limits. The burning question is: how do you spend the hours of darkness? With what the engineer might describe as a 'refuelling system', the animal or bird must keep the 'essential services' ticking over, even though it may be asleep. (A bird does not sleep perched on a branch without a built-in automatic control system controlling the grip and balance.) The digestive systems of mammals and birds are as subject to the rules of size at least as much as are the wings of a bird to fly or the legs of a mammoth to stand. The rule that emerges is that the smaller the creature, the more frequently it must feed. If it cannot survive a 12-hour fast, it cannot exist.

But the same *kind* of rule must surely apply due to the earth's rotation around the sun, which produces the *seasons*. In a snow-covered, frozen land the smaller vegetarian mammals, the caterpillars and other creatures, must surely starve? Nature solves this problem through the mechanism of hibernation, an experience which we cannot share. If a living creature cannot eat, it must die. Hibernation is a kind of 'living death' where many of the normal functions of the body become totally inhibited. Butterflies and moths manage to spend the British winter in suspended animation at all stages of their lives. One species will winter in the egg, another as a hibernating caterpillar, a large number of species winter as a pupa, whilst in others, the adult female finds a warm airing cupboard in which to hide (in which case how does she know when the first warm 'spring' day occurs prematurely in February? – for she undoubtedly does and emerges to lay eggs outdoors).

The pupa stage of a butterfly is also a hibernation that is necessary for quite different reasons. Insects in general, arachnids and snakes grow by a much larger factor than do mammals and birds. These latter manage with one skin that can stretch, or have its cells replaced at a rate that can cope with a size increase from perhaps 7 lb at birth to 210 lb fully grown (in the case of *H. sapiens*). The factor of 30 in volume amounts only to $\sqrt[3]{30} = 3.11$ in linear dimension. Each egg laid by a female moth must contain a complete caterpillar which must grow to be larger than the body of its mother (in the case of a female), and the linear-dimensional increase is of the order of 50 (125 000 the volume). Skins that cannot cope with such growth, in a matter of a few weeks in the case of some caterpillars, must be shed from time to time. The rapid growth also demands rapid eating, and caterpillars have powerful jaws (mandibles) with which they systematically demolish green leaves by moving their heads so as to cut successive concentric annuli. But the adult insect has wings and much longer legs and, for a reason that escapes us, it feeds only by sipping nectar through a living drinking straw (the proboscis) which is coiled up under the head when not in use.

Now, a creature that has to change its mouth parts cannot eat during the change, so, like creatures passing the long winters, it too must 'die' for a time. That some caterpillars make the change to a moth in a fortnight is still one of life's wonders, for me. (The common Vapourer moth, which has almost become a city-dweller in the UK, can do this.) But a creature that only requires a set of wings and a new set of legs and retains its mandibles only passes through a 'nymph' stage where it is encased in its own 'plastic' cover but does not hibernate. The dragonflies are classic examples of this form of life history.

4

Size is everything

I suppose I was first alerted to the importance of size by the physics book that I used at school that I referred to on p. 40. Entitled *Properties of Matter*, it dealt with all the 'left-overs' from the other books on heat, light etc. Topics such as the mechanical strength of materials, surface tension in liquids, the viscosity of both liquids and gases and similar most important concepts were included. Although there was no biology in this book, one particular sentence read: 'If a man and an earwig both fall over a cliff, the man falls to his death, the earwig floats down unharmed. But if instead both are soaked with water, the man will suffer little ill-effect but the earwig will most certainly drown.'

The subject matter to follow was, of course, the viscosity of air on the one hand and the surface tension of water on the other. I have since heard the first part of the statement amplified by means of a series of statements relating to the fate of various creatures that were dropped down a deep mine shaft, thus:*

> a spider simply wonders why it has gone dark!
>
> a mouse waits patiently for the slight jolt at the bottom and then runs off;
>
> a domestic cat lands on its feet and survives;
>
> an Alsatian dog is killed;
>
> a person is broken;

* Buller points out that the spores of the fungus *Collybia*, which measure 5 microns × 3 microns (0.0002 × 0.00012 inches), have a terminal velocity of 0.5 mm per second (= 0.001 mph).[1]

a Clydesdale horse is smashed;

an elephant explodes!

Chapter 3 referred briefly to 'terminal velocity' in free fall, which is the all-important quantity in these considerations. Only when an object falls under gravity in a perfect vacuum does it continue to have constant acceleration. In air it encounters atoms by the billion billion per second and this produces a similar kind of drag force to the one experienced if a plate is drawn through water with its face pointing in the direction of motion. The drag force is proportional to the *area* at right-angles to the motion and to a complex function of velocity which may be appreciated simply by saying that at low speeds (20 mph) it increases linearly with speed. At speeds up to three times this, the drag force rises with (speed)^2. At hundreds of miles per hour it rises as (speed)^3, and so on.

Now, the force of gravity on a body depends on the product of its density and its volume. Most animals have a mean density between 0.9 and 1.0 g/cc (900–1000 kg/m^3 or 62.5 lb/cu. ft), which means that all will just float in water and therefore have the ability to swim.* So when we study terminal velocity in free fall, the general rule that resistance is increasing with area (length)^2 whilst driving force (gravitational) increases with volume (length)^3 means that larger bodies will have higher terminal velocities.

The second mental jolt that I received on this topic occurred when I was acting as demonstrator for the late Professor Sir Frederic Williams as he was lecturing to first year undergraduates at Manchester in 1951. During the course of one lecture he proved conclusively that if humans had been 1 inch high when fully grown and all properties of matter had stayed the same, electrical engineering as we know it would not have existed. Such electricity as could be used would have had to be provided from chemical batteries and would have cost, on 1951 prices, 15 shillings and 9 pence per unit! This was to send me off in search of the goodness factor for electrical machines to which I referred in Chapter 3 (pp. 54–65).

---

* The only mammal that is incapable of swimming, not because of its density or shape but because it never had a need to do so, is a camel.

The cause of shape

One may subscribe to the belief (expressed in Chapter 6) that whilst topology is the theme that runs through the whole of the story of life and of engineering, this neither precludes the existence of an overall cause of, nor attempts to explain the reasons underlying, the amazing variety of shapes both animate and inanimate. One person[2] sees it all as the result of the two dominant fluids on earth, air and water (pp. 149–151). This puts it all in terms of the substances peculiar to our planet. Another[3] sees it as an elaborate extension of Euclidean geometry in three dimensions (pp. 151–155). Yet a third[4] goes further and declares it to be the result of the limitations imposed by space itself, perhaps the nearest a human has ever come to putting the mysteries of relativity and the quantities in which it deals, gravitation, inertia, electrostatics and electromagnetics, into physical ideas that most of us might be able to grasp (p. 147 *et seq.*). One of the powerful aspects of this last approach is that it suggests a reason why absolute size may have a meaning and is not therefore merely a relative thing.

Meanwhile, there is a temptation to extend the philosophy to more things than shape. Many scientists have seen the model of an atom as a miniature replica of a solar system (p. 70). Electrons are the counterpart of planets. The solar system might then be but one atom in a complex super-molecule (the Milky Way). That galaxy is then one molecule in a lump of 'super-matter' that we call the universe. Alas, that would be all too easy. Besides, it would imply a near infinity of other universes, and even though such *may* exist, they would not do so for such a reason as to maintain a basic structure so simple that mere humans could understand all of it.

A journey through size

Let us take a pictorial journey through size, notice the *apparent* repetition of shape and then see the subtle differences between things that superficially appear to be identical.

We begin with a tiny white speck that most of us would call a star, and needs no illustration. But when viewed through a powerful telescope it is seen to be a collection of white specks of a variety of sizes (Fig. 4.1). They appear to form the shape of a 'frozen' vortex. Our guide on this journey tells us that the shape is indeed the result of rotation, but at a rate so slow (in revolutions per minute) that we shall not be conscious of its movement in our lifetime.

Our camera moves in. It sees one of the bright specks as a flaming ball that spins on its own axis, and this in turn is attended by numerous other balls, not themselves ablaze, but visible to us because they reflect the light of the centre ball. Each of the cool spheres also spins on its own axis. Their movements around the centre appear to be circular orbits but closer examination shows the path of each to be an ellipse in a plane with the fiery sphere at one focus. Is it not already a source of wonder that such a system is so ordered that the geometry of one planet's motion is a mere 2-D affair, describable to great accuracy in terms of a single second order algebraic equation? Did the Creator *also* do co-ordinate geometry in his schooldays?

A super camera now focuses on one of the cool, orbiting spheres. It is surrounded by masses of visible gas but there are adequate gaps between the clouds for us to see the solid surface below. We look through a space at increased magnification. It appears to consist of two kinds of surface; the one looks quite smooth, the other displays shadows that indicate lumps and hollows. The sphere is neither quite perfect geometrically, nor perfectly smooth. We pick out a lumpy portion and magnify it still further. It is extremely varied and picturesque. For the first time we see some straight lines. Our guide tells us that they all began as the result of a Being called Euclid. (Until this point everything has been curved and spinning.) We can just see that some tiny specks on the surface are moving. We move in closer to see what they are. They are making a new straight line along the surface. The creatures are of several types. Some are angular with many flat surfaces, the others more 'natural', i.e. bounded entirely by curved surfaces, the only 'orderliness' of their topology being a bilateral symmetry. Our guide tells us that this is entirely due to

Fig. 4.1. The helical form of a galaxy.

their need to travel about the surfaces on which they live.*

We focus on the surface skin of one of the moving creatures. Not surprisingly perhaps, it is not smooth. It consists of a multitude of depressions and of stalks protruding, one from each. Let us go inside the surface and examine the substance in still greater detail. (We fit a microscope to our camera.) We find this substance to be of great complexity. Our guide tells us that increased magnification will show us cross-like structures which are characteristic of the non-angular creatures that moved (Fig. 4.2a).

At this point let us invent a size unit that will help us to appreciate the extent of our journey through size. In journeys through time we speak of millions of years, where a year is measured as the time it takes our planet to orbit the sun. For longer times we tend to use non-quantitative words like 'aeons' and 'ages'. We even recognised time contraction through Western religious writings long before authors were said to write 'science fiction':

> A thousand ages in Thy sight
> Are like an evening gone.
> (*Hymns, Ancient and Modern*)

In a journey through distance we measure in kilometres (to please Charles de Gaulle) until the numbers again become unmanageable. (10^{20} kilometres conveys little meaning when there are only 10^{89} particles in the universe.) Then we use the distance travelled by light in one earth year as the yardstick of space. Let us therefore invent a new size relationship as enormous to human minds as the 'age' or the 'light year'. Let the scale factor be 10 000 000 000, or 10^{10}, and let us then call it a 'decem' (for fairly obvious reasons, like pleasing Charles de Gaulle *and* Julius Caesar!). Two decems are to be $(10^{10})^2 = 10^{20}$, and so on. As an example of the size of a decem, the distance from the earth to the moon is just over 10^{10} inches and 200 000 mph is just over 10^{10} inches per hour.

Returning to the cross-like structure of Fig. 4.2a, our guide assures us that this *is* the vital ingredient of the smooth creatures that moved almost a decem away. From now on we must rely on

---

* Martin Gardner – see p. 135.

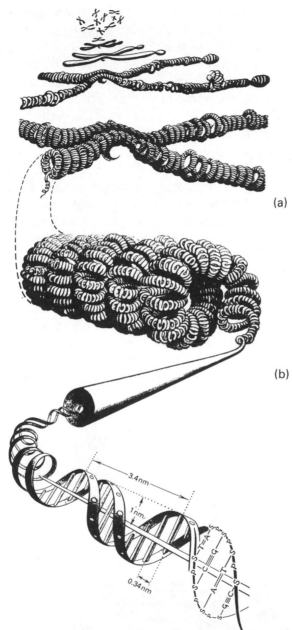

(a)

(b)

3.4nm

1nm

0.34nm

Fig. 4.2. An artist's impression of continuous magnification from a chromosome to DNA.

an artist to continue our journey, for we have no instruments that
will magnify any further, but there are good reasons for believing
that the infrastructure of the crosses is as shown in Fig. 4.2b.
The subsequent expansion of this structure leads us to a double
helical shape of what our guide tells us is a 'complex molecule'.
It is at this point that some of us may reflect that 'this is where
we came in' (as people used to say in continuous performance
cinemas), for the nebula of Fig. 4.1 was a helix also. But we
return to our guide and ask what a molecule is. We are told that
it is something like the model shown in Fig. 4.3, but that we
must never assume that these are anything more than models,
and that the stalks that join together the blobs are mainly in the
mind! We hazard a guess that if we could really *see* the blobs they
would not be perfect spheres, nor would they be smooth. It is a
far cry even from *that*! Any one of the blobs is called an atom
and a model of an atom looks like the object shown in Fig. 4.4.
Again we are inclined to make comparison between this model
and what we saw earlier nearly 2 decems larger. There is a central
'sun' and 'planets'. But now the orbits are not roughly in a plane.
This is a similar sort of difference to that between Fig. 4.1 where
the helix was in a plane and Fig. 4.2b where it was a 3-D helix.
There are other differences. Some of the atomic planets orbit at

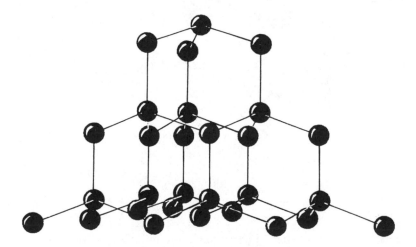

Fig. 4.3. A complex molecule.

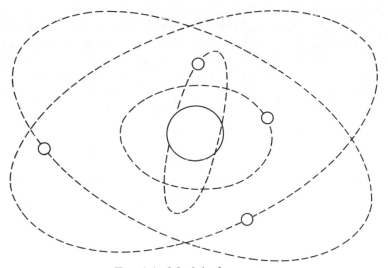

Fig. 4.4. Model of an atom.

the same radius from the centre. Our guide tells us that there are hard and fast rules about this. There can be atoms with either one or two planets. They will be at the same radius, i.e. they will orbit around the surface of an imaginary sphere. But when a third is present it must go out to a bigger radius. The inner 'shell' is *full*, we are told. The second sphere can contain up to eight and then a third, larger, shell is needed. The 'sun' is not fiery. The resemblance to the fiery sun is only superficial. **But** – again we guess that if we could really see the little spheres they would be neither perfect nor smooth – worse than that, *they* might not even be solid either!

Our guide says that the central body is called the nucleus and that it is made up of 'protons' and 'neutrons'. The orbiting planets are called electrons and their mass is only 1/1845 of that of a proton or neutron. The latter are made up of quarks held together by *gluons*, at which point we assume that our guide is pulling our legs. 'Stuck together by glue-ons.' Our guide is offended. He takes himself *very* seriously, but his imagination for the physical universe seems to have let him down when it came to inventing new words for new things.

By the way, who is our guide?

In later chapters I shall discuss 'collective thinking' in the context that colonies of 'social insects', such as bees, wasps and ants, and swarms of butterflies, locusts or even birds, appear from many points of view to behave as if they were just one living organism of much greater complexity – greater perhaps even than *Homo sapiens*? Our guide is such an entity. It is the sum total of human experience since the dawn of civilisation that embraces logic, mathematics, science and engineering and is collected together in a vast library of books (which library, of course, is distributed in buildings throughout the world). Teachers of science strive to simplify their teaching by trying to extract the apparently more *fundamental* bits and *fundamental* rules. And progress is the everlasting search for the more fundamental.

Our guide therefore has its limitations too. It sends out scouts ahead to explore what might be accepted as fact in the next century. I have seen matter broken down by some authors beyond the quark and the gluon, and for me quite the most fascinating (and for an engineer, perhaps the most plausible) is contained in a paper in the journal *Speculations in Science and Technology*.[5] (Only 'speculations' are published here. Perhaps it is not surprising that within such pages there should be a paper on the structure of the electron, nor that the latter's components are said to form 3-D helices! We have, after all, moved through another decem since the DNA molecule.)

Paraphrasing the paper by Gulko, space is assumed to be full of fluid (without ever calling it 'ether') that exists 'as a residue of the energy released at the creation of the universe', i.e. energy as a fluid. Within the fluid there are vortices (another form of helix but like a doughnut) in which the fluid circulates as shown in Fig. 4.5a. It is, of course, exactly like a smoke ring and there itself is a phenomenon we can both see and explain, and yet is surely a thing of beauty and a source of wonder.* Gulko proposed

---

* A small digression is probably permissible. One of the finest physical demonstrations I have ever seen was devised and performed by the late Sir Lawrence Bragg, at the Royal Institution in London. At the end of a long bench he had a smoke ring generator constructed as shown in cross-section in Fig. 4.6. A is a wooden box in which one end, B, has been replaced by a sheet of parchment stretched tightly over the wood. The opposite end of the box carries a circular hole, C. D is a pendulum bob hanging from a string and in contact with the parch-

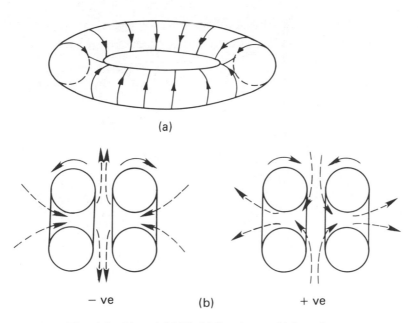

(a)

− ve (b) + ve

Fig. 4.5. (a) and (b) Fluid flow in toroidal vortices.

that a *positive charge* be defined as a fluid density higher than its surroundings, whilst a negative charge was a region of lower density. The density variations were caused by pairs of vortices pumping the fluid axially inwards or outwards as in Fig. 4.5b, where the 'doughnuts' are seen in cross-section. This explains

ment. When the box lid, E, is lifted, smoke can be blown into the box and the lid replaced. The pendulum bob is now drawn back to the position shown dotted, and released. When it strikes the parchment, a 'standard' ring at a known velocity will be ejected each time.

Sir Lawrence sent a smoke ring travelling along the bench and, unknown to his audience, he had started a stop watch at the instant the pendulum bob hit the parchment. At the far end of the bench, his assistant, Bill Coates, adjusted the position of a lighted candle so that the rim of the oncoming smoke ring would pass through the flame. As it did so, the air currents in the vortex blew out the flame and Sir Lawrence stopped his watch. 'Now', he said, 'I will repeat the experiment *without* the smoke.'

He drew back the pendulum on an empty box, started his watch secretly and 'watched' an invisible vortex travel at the same speed towards the candle. At the appropriate time he counted down: 'Three, two, one – out' and out went the flame – magic!

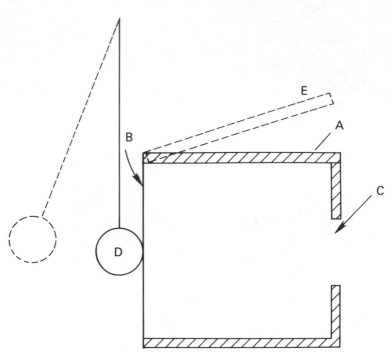

Fig. 4.6. Apparatus for producing smoke rings.

the attraction/repulsion of charges simply as the movement of fluid from high to low pressure areas.

Now, a single vortex in the fluid he suggests 'is a candidate to constitute a neutrino'. And a neutrino, as any good physicist will tell you, is definitely 'matter'. So the chain is complete from the 'real' (matter) to the 'mystic' (charge, the hitherto inexplicable). Gulko's paper may be fantasy and may be classed as 'crackpot' by reputable scientists, but I recall a report of a meeting in New York in which Neils Bohr rose to comment on a paper that had just been presented by Pauli. 'We are all agreed' said Bohr, 'that what you are saying is crazy. The point that divides us is whether or not it is sufficiently crazy to have a chance of being correct.'

It has been a strange journey through size and we have covered more than three decems. It is stranger than the best of science fiction. It remains an enigma. Things that looked alike were not *quite* alike. That there was an *order* in it all is not in doubt. That this order is not clear to us is not in doubt either. One conclusion emerges loud and clear. In a journey through size, as in journeys through space or through time, we see ourselves 'in the middle', and the nature of space, the 'nature of things' (Sir William Bragg) appears to change only 'at the ends'. There is a very nasty, deep-rooted feeling that somewhere we have deceived ourselves with our science and our logic. We can't have been *that* lucky always to be just in the middle.

The light pen

Now, modern electronics can help us out just a little in unravelling this enormous paradox. Its origin was a subject as earthy as that of engineering graphics. Draughtsmanship has certainly moved with the times. There are two electronic forms of visual display, other than the cathode ray tube that has long been dear to most of us, one might even call it a member of the family.* These are the light-emitting diode (LED) and the liquid crystal (see p. 252), both of which are reproducible on the same microscopic scale as is the phosphor on the cathode ray tube. This is not to say that it is infinitely fine. It is comparable with the grain size in photographic materials.

With a screen of electro-luminous material available, the draughtsman uses a 'light pen'. This is a charged probe (which looks just like a ball-point) with which he merely indicates the location on the screen where he wishes this or that shape to be drawn. He brings the pen point very close to the surface but does not necessarily have to touch it. What shape is drawn has been previously decided by his pressing keys on a display not unlike a typewriter or desk calculator. The great thing is that he is relieved of the care of ensuring that his lines are straight, circular, or whatever. If he wants a straight line from A to B, he indicates A with the pen, moves it to B, and the straight line appears no

---

* Some would say an 'over-dominant member'!

matter how wiggly was the way the actual pen moved from A to
B. He had *demanded* a straight line from the keyboard.

The number of facilities that such a system provides is staggering, for the electronic circuits have 'memories'. Thus, among
other things, the screen can reproduce your drawing increased
or decreased in size, and you can add further detail at any scale.
The operator can show you an aeroplane that he has drawn, for
example, in side view. With the light pen he indicates the tail fin
and presses a button marked '×10' (times 10). The whole tail
fin fills the screen and, among the detail he had previously
included at this size, you can see a motif that appears in the
middle of the fin. In the motif is a reference number too small
to see at this scale. He touches the motif and calls again for
'×10'. Now the motif fills the screen. We read a 10-digit number
now an inch long. He can now erase or change any of these digits
at will with the light pen/eraser.

He tells us that in the memory banks he has drawn the aircraft
in relation to its position in the solar system and proceeds to
demonstrate that he has. Calling for '÷10' twice brings us back
to the whole aircraft. 'Divide 10' again shows the aircraft much
smaller but nothing else. 'It is in flight,' he tells us, and another
'÷10' shows the ground, houses, a church, etc. The aircraft
is flying at 3000 feet (900 m). Another '÷10', and the aircraft is
little more than a short line, the earth is a blurred line that is
noticeably curved – we become conscious of the curvature of the
earth. Two more '÷10' operations and we see the whole earth.
The aircraft is a bright spot on its surface.* Four more '÷10'
instructions, and we see the sun and a bright dot at the other
side of the screen. The dot is not the earth, it is the aircraft!
Within the dot is all the detail including that reference number,
not in the material of the viewing screen of course, but in the
machine as a whole, i.e. in its data banks. From reference number
to solar display, the machine has come through a decem in size

---

* Even the early 'astronauts' (Gagarin, Shepherd, Grissom, Glenn,
Carpenter) were less than a millimetre (0.04 inch) off a 30-cm diameter (1-foot)
geographer's globe. Borman, Anders and Lovell were the first men really to go
into space. But the first true astronaut who literally flies among the stars may
already have been born!

for just 10 touches of the button. It is the ideal machine for making journeys through size.

The ends and the middle

If then we insist that we really are in the middle of a journey through size, what might we reasonably expect to find at the ends, ignoring what our guide told us earlier about the change in nature? There are many examples in living things from which we can learn. The almost inevitable answer of course is: 'You ask a meaningless question.' Take for example the question: 'Where does a river start?' Apart from those which emerge as spring water from the ground, the correct answer to this question is: 'Only in an atlas.' Generally rivers start in a marsh which gradually gets less muddy and more watery as you go towards the sea. There is no one point where you can say that the water changes from totally stagnant to continuously flowing. Nor will there be a point where you can say that water has emerged from thick mud.

There is such a marsh in many facets of human endeavour. It is modern technology that has often made a nonsense of such things as world records. Before the days of the jet engine there was a real meaning to a land speed record and a water speed record. But once the propulsion has ceased to come from the land or from the water it would appear that any aircraft touching land or water *once* within the measured mile by means of any sort of projection below the aircraft qualifies for the record book. And what of someone orbiting the earth in a few hours? What is even more obvious is that before space flight, an altitude record by balloon was a meaningful target, but after moon landings an altitude record becomes meaningless in the face of the question: 'Height above what?'

Absolute size

It has already been pointed out (pp. 48–49) that when time is speeded up, small objects have a habit of appearing larger. In the world of engineering the precise relationship between time and size is different in different situations. In the case of electric motors, for example, reducing all linear dimensions of one particular motor design by a factor n means that the supply frequency (which is dimensionally the reciprocal of time) must be multiplied by n^2 if the scaled-down version is to be as good as the model (in almost every interpretation of the word 'good').

In relativity and other advanced concepts, there is a real danger in going all the way with what is called the 'Space–Time Continuum'. Absolute size may have a meaning. If, for example, a man's fully grown height was reduced to $(1/10)^{18}$ (nearly 2 decems) of what it is now, and if an atom really *was* like an orange, orbited at miles distant by electrons the size of pin heads, and we lived on an electron which was a small rotating sphere, we would know with certainty (from our guide) that we were held on to our 'earth' by electrostatic forces, and that our earth was held in orbit around its sun by electrostatic forces also. Gravitational forces would be so trivial that their use would be relegated to evening parlour tricks such as we now perform for small children by rubbing a plastic rod on fur and picking up bits of paper – by electrostatics!

So if gravitation and electrostatics exchange roles over a range of nearly 2 decems, we can only draw one of two conclusions. Either, there is a meaning to absolute size, or there is a unique connection between gravitation and electrostatics that has so far eluded us. Perhaps they may be different facets of one whole, like two sides of a twisted ribbon.

When we examine ferromagnetism, we find a basic phenomenon connected with electron spins in outer shells of the atoms of seven particular elements, the commonest of which is iron. In these seven elements there is an out-of-balance spin in the outer shell, and atoms that have such imbalance try to align themselves so that the spin axes lie parallel. It is the effect of this out-of-balance spin that we call 'magnetism'.

In the first part of this century the general view of ferromagnetism was that an unmagnetised bar of iron consisted of billions of tiny magnets lying higgledy-piggledy in all directions. When you magnetised it, for example, by stroking it with a magnet, you tended to make the tiny magnets line up but it took work to do this, and it did not all happen at once, so you had to keep repeating the process in an attempt to make them *all* line up, in which condition it could not be magnetised any further and was said to be 'saturated'. But work at the Bell Telephone Laboratories in the USA in the 1930s showed that neighbouring atoms pulled each other into line without external interference, so that iron was, so to speak, self-saturating. How then could you have unmagnetised, or partially magnetised iron? It emerged that the iron divided itself into regions which became known as 'domains' and it was the arrangement of these domains which gave the external impression of a bar of iron being magnetised or not.

One of the most far-reaching rules of science appears to be that all things tend to a position of minimum energy. When you lift an object upwards you move it away from the earth and in doing so you give it energy. But, given the chance, it will return to the earth and thereby reduce its energy. (The fact that a person lying down requires less energy than when sitting down and still less than when standing up is not an example of this scientific principle but, as an analogy, perhaps it helps!)

If we could examine the pattern of domains* in an unmagnetised iron crystal we might find them to be as shown in Fig. 4.7a. The boundaries between domains were called 'Bloch walls' and it costs energy to set up such a wall. When iron is magnetised it changes in length very slightly, a phenomenon known as magnetostriction. This produces strain (magnetostrictive) energy within the iron. If the iron sends magnetic field out into the space outside itself, this costs still more energy. This is called 'magnetostatic' energy.

Any given piece of iron will arrange its domains and their walls so that the **total** energy (the sum of Bloch wall, magnetostrictive

---

* The workers at the Bell Laboratories developed a polishing and etching technique that allowed them to do just this.

and magnetostatic) is a minimum. So how can we have de-magnetised iron?

Figure 4.7 shows cross-sections through differently sized iron crystals. In a large crystal, as at (a), it can be shown that a minimum energy situation, and therefore a stable system, involves the domains in arranging themselves as shown. No magnetic field 'spills' outside the crystal (i.e. there is no magnetostatic energy) so we say it is unmagnetised. When an external field is applied, the domain boundaries move to a position as shown in (b). The crystal is now magnetised because there are bigger domains pointing in one direction than the other. Much smaller crystals are stable in the condition shown in (c), in which less energy is expended in driving magnetic field outside the ends, in loops in space, than in the combined strain energy in the crystal itself and the energy needed to support the domain walls. Still smaller crystals, as fine as flour, are stable with no domain walls, as at (d). Discovery of this fact in the 1930s started a whole new industry of 'powder-metallurgy' where very powerful permanent magnets are made by 'sintering' (heating almost to melting point) ferrous powder immersed in a strong unidirectional field until the powder solidifies into a block.

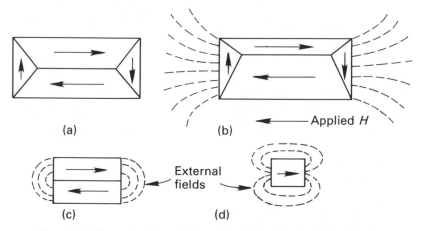

Fig. 4.7. Magnetic domain boundaries in iron crystals. (a) An unmagnetised specimen. (b) The result of an externally applied field. (c) Stable condition in a very small crystal. (d) Minute crystals are self-magnetising.

So it was a size effect entirely. There are definite sizes at which the magnetic patterns 'flip' from one state to another. Can this be a size milestone from which we can measure all other sizes? Again, it all depends on whether there is a connection between spin and magnetism that has eluded us. Perhaps things have always come in fours, like the 'earth, fire, air and water' concept of the Ancient Greeks. In which case, the space–time continuum might be replaced by the 3-D model shown in Fig. 4.8, and how the four forces are co-related depends on where you are in the 'size position' around the four-sided band.

A further comment on absolute size and the effects of gravity on smaller things comes from the journal *Speculations in Science and Technology*, to which reference was made on p. 72. James D. Edmonds writes:[6]

> Time is relative in *just such a way* that all inertial observers are equally 'at rest'. This is one of his (Einstein's) two basic postulates. The other is that no inertial frame is special which requires that c = c' for the speed of light in every Lorentz frame. This is very pretty except that *no Lorentz frames really*

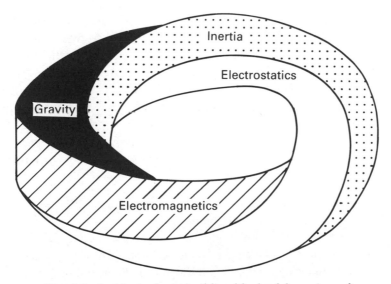

Fig. 4.8. Is this the basic building block of the universe?

exist in nature! Space is *curved*. Neglecting curvature means doing sloppy measurements or, alternatively, doing *accurate* measurements in *very small* regions of space–time where the curvature can sometimes be neglected. But this is ridiculous since humans, and nature generally, are NOT scale invariant. There is a fundamental length in nature. Atoms have a definite size and indeed *they curve the space around them*! . . . Suppose G were 10^{40} times stronger than it is. *Nothing in principle* would be changed about relativity or its basic postulates. But then you could not do any experiment without worrying about the curvature produced by your experiment.

Characteristic length and characteristic time

Living things have a relativity all of their own. In mammals, for example, a 10-ton elephant has a gestation period of 760 days. A $\frac{1}{20}$ ton human carries her offspring for only 270 days. The ratio of gestation period is 2.8, so the relationship with distance seems to be such that time varies as the fifth root of volume or $2.8^5 \approx 200$, so $(\text{time})^5$ is proportional to $(\text{length})^3$, i.e. time $\propto (\text{length})^{0.6}$. (Remember that in electric motors, time $\propto (\text{length})^2$.) Midges have had their wing beat rate measured as 2200/second. Their wingspan is about 5 mm ($\frac{1}{5}$ inch). A bird with a 60 cm (24 inches) wingspan makes about one beat per second. So a reduction factor of 120 in length reduces time by 2200 times, suggesting an index of about 1.66, which perhaps warrants another look.

Edward S. Taylor's fascinating book *Dimensional Analysis for Engineers*,[7] in a section entitled 'Geometric similarity in nature' begins:

It is easy enough for man to make an unsuccessful design – and in the initial stages of the design process it is often impossible to be sure that a design is indeed workable. On the other hand, if we observe biological organisms we see only successful designs, all others having been eliminated by the evolutionary process.

He goes on to say:

> While a hummingbird does not look or act like a whale, the
> nearly one-to-one correspondence of vital organs and bones of
> these two vertebrates is astonishing indeed. In order to find
> out the effect of size it is well to study a large range of sizes
> so that the effect of changes in size will be large enough to
> swamp the effect of changes in form. Fortunately nature
> provides an enormous range of sizes.

Just how enormous this is can be seen from the following table
of weight ranges, set out according to the major families and
structures of living creatures.*

Seeds	2.2×10^{10}
Mammals	1.9×10^{8}
Mammals (excluding whales)	2.0×10^{7}
Birds	7.8×10^{4}
Birds (including prehistoric)	2.3×10^{5}
Snakes	7.8×10^{5}
Insects	2.0×10^{7}
Flowers	5.8×10^{9}

and then

Electric motors and generators	6.2×10^{8}

Some interesting maxima and minima which I discovered in
compiling the above table are as follows.

> The tallest tree, a Redwood, is 111 m (366 feet) high.
> Lifting water to such a height against gravity is no mean
> feat.

> The smallest seed, that of an Epiphytic orchid, weighs
> 1/1 200 000 g (1/34 000 000 ounce).

> Termites have been known to build a column whose height
> is 2500 times the length of a termite. If we could build a
> tower in the same proportion it would be 4600 m (15 000
> feet) high (half the height of Mount Everest).

* Compiled from the *Guinness Book of Records*, 27th edn (1981). London:
Guinness Superlatives.

A prehistoric Tanzanian Brachiosaurus has been measured as 22 m (74.5 feet) long, 6.4 m (21 feet) high and its estimated weight is 78 tonnes. It has been suggested that other specimens of this creature might have reached 27.4 m (90 feet) in length and topped 100 tonnes.

The prehistoric flying Pteranodon had a wingspan of 8.2 m (27 feet) and is estimated to have weighed only 18 kg (40 lb).

The sea-water Protozoa known as *Dinoflagallata gymnodinium breve* lives at a density of 500 000 000 per litre (2300 000 000 per gallon).

Bacteria have been discovered over 40 km (25 miles) above the earth's surface. The smallest known virus is estimated to measure about one millionth of an inch (1/40 000 mm) in its longest dimension.

The largest flower is that of the 'stinking corpse lily', measuring nearly a metre (3 feet) in diameter, weighing 6.8 kg (15 lb) and apparently smelling 10^6 units of obnoxiousness!

Whales have been known to travel 8000 km (5000 miles) in three months.

A flying fish has reached a height of 180 m (600 feet). Vultures, using thermal air currents for soaring, have been recorded at 1830 m (6000 feet). Also using air currents, Monarch butterflies have crossed the Atlantic in three days (an average speed of over 18 m/s (40 mph)).

Some of the feats of birds are perhaps the most remarkable of all. A wheatear makes the trip from Greenland to Spain in 48 hours and loses one-half of its body weight in the process. The Arctic Tern flies from Alaska to Antarctica and back, making the round trip of over 64 000 km (40 000 miles) in three months, an average speed approaching 9 m/s (20 mph), whilst a swift is thought to be capable of continuous flight for three years. These feats are only possible, of course, because of the available food. The sea birds can dive for fish en route and the swift collects all its food from the flying insect population. Long-eared owls can hit targets which are illuminated at the intensity provided by a candle flame 356 m (1170 feet) away. A golden eagle can see a 46 cm (18 inch) target at 3.2 km (2 miles). This means with a resolution of 0.0000025 of a degree of arc (≈ 0.01 second of arc). This is of the order of 50 to 100 times the resolving power of the human eye. In 1840 there were estimated to be 9 000 000 passenger pigeons alive on earth. By 1914 there were *none*! Stuffed specimens fetch hundreds of pounds at auction rooms.

In the species *Searches holboeki* (a deep-sea angler fish), the female is 500 000 times the weight of the male.

The largest creature of all is the giant jellyfish, *Cyanea artica*, with a tentacle span of 75 m (245 feet). Considering that the smallest living thing to date is thought to be a mycoplasmic organism 0.0001 mm (0.000004 inches) in diameter, the overall length ratio in all living and moving things is of the order of 10^9, which implies that they might have a volume and weight ratio of 10^{27}, but of course both the extreme creatures are far from solid.

In such a spread of size and form it would surely be surprising if any sort of rule vaguely resembling a scientific 'law' were to emerge!

Taylor defined 'characteristic time' as the time which it takes a creature using oxygen as vital life support to consume a quantity of oxygen equal to its own mass. 'Characteristic length' he defined simply as the cube root of the mass of the animal. Taylor then remarked: 'If vertebrates were limited by the same laws as internal combustion engines (car engines, aircraft jet engines, etc.), we would expect the characteristic time to be proportional to characteristic length.'

He proceeded to attempt to relate characteristic time to characteristic length, but the relationship was complex, involving logarithms, and was in no way exact. However, we saw in Chapter 3 that for electrical machines, l^2 must vary inversely with time and this does not fit Taylor's more complex growth relationship, presumably because the living creature is a much more complex structure. On the other hand, both Taylor and D'Arcy Thompson make the point that the frequency of wing beats in both birds and insects varies inversely as the wing span. In this case a linear relationship between characteristic length and characteristic time means a constant characteristic *velocity* of wing tip for all sizes of creature of the same fundamental shape. One suspects the all-powerful value of the Reynolds number again, for no bird would want turbulent flow around the fastest-moving feathers. A humming bird approaches frequencies of the order of 75 beats per second. Slow-motion photography shows the motion of each wing to be such that the tip traverses nearly a 180° arc with 'natural', i.e. sinusoidal, acceleration and deceleration. Hence, if the radius is $\frac{1}{8}$ foot, the maximum velocity is given by:

$$\frac{1}{8}\left[\frac{\pi}{2} \times 2\,\pi \times 75\right] \text{ feet per second}$$

which is 93 feet per second or 63 mph (28 m/s). To a good cricketer, this calculation might prompt him to exclaim, 'seam bowling!' (as practised to a fine art by such 'greats' as Alec Bedser). In seam bowling the bowler polishes one side of the ball only. He then delivers it, without spin, so that the stitching around the diametrical seam lies at the correct angle to the velocity direction, and when he gets the speed right (and the tolerance is only a few per cent) the Reynolds number shows that laminar flow obtains on one side of the seam and turbulence on the other. This causes the ball to move in a curved path, to the considerable discomfort of the batsman. In humid atmospheres the effect is increased. A weather forecast in the bird world would be quite different from those we hear on radio. It would include such information as, 'The viscosity in the Outer Hebrides is 0.00018* and the humidity 65%.'

If humankind was 'planted' on earth by an extra-galactic crea-

---

* The coefficient of viscosity is the tangential force per unit area needed to maintain unit relative velocity between parallel planes unit distance apart. Its units are therefore force × time/area.

ture, then whoever it was must have had a sense of humour, for we find natural phenomena that just 'do not fit' any theory that puts things in order, either in terms of physical laws or of classification. One needs only examine the duck-billed platypus for evidence of this! But there are other, less well-known practical jokes of this kind. R. L. Gregory and E. H. Gombrich, in *Illusion in Nature and Art*, have pointed out the pupa of a moth that displays a picture of a monkey's face.[8] The lantern fly of the Far East mimics an alligator, but the fly is only 7.6 cm (3 inches) long and the whole of its proboscis lights up at night! In neither of these cases is the mimic in any way likely to be mistaken for the original because of the enormous size disparity.

Now, the largest seed in the world is another matter. Native to the Seychelles islands, it is a species of coconut whose leaves and fruit are suspended over the sea. The nuts, including husks, are very large and, when ripe, they fall into the sea and are washed ashore. They are known as 'coco-de-mer'. But their shape! (Fig. 4.9). They are complete replicas of the torsos of human females from just below the navel to the tops of the legs. The size is the same, the 'pubic' hair is all properly located, the buttocks are smooth. You cannot examine one of these seeds without mentally thinking something like, 'Does God *mock* us?' Even within the millions of species of living creature, the likelihood of a coincidence of this kind between parts of species as remote as woman and tree is, to say the least, unconvincing.

Euclid and Euler

Most plants and animals have only a superficial resemblance to circles, cylinders, cubes and right circular cones. Even the earth itself is not a perfect sphere. Just occasionally, we see what appears to be an isolated case of Euclidean order, for example, the hexagonal structure of honeycombs. We are inclined to credit the bee with being more 'civilised' for having invented it – until we look through a *microscope*.

There we see the straight lines and equation-regulated curves

Fig. 4.9. The largest of all seeds – the coco-de-mer.

of our academic years, solid figures not only of regularity, but of
great artistic beauty. Bilateral symmetry gives way to spherical symmetry with planes, sharp edges and sharp corners no longer pro-
hibited. God *must* be a mathematician, but He, apparently, could
only do it up to a certain size.

The great mathematician Leonhard Euler went on working
profitably well into his eighties, using servants to read and write for
him after he lost his sight. One of his famous theorems concerned
solids bounded entirely by planes. The simplest of these is the tetrahedron (Fig. 4.10a) bounded by four triangles. It can be seen to
have four faces (areas), six edges (lines) and four vertices (points).
If a corner is sliced off by another plane, as in (b), the resulting solid
has gained:

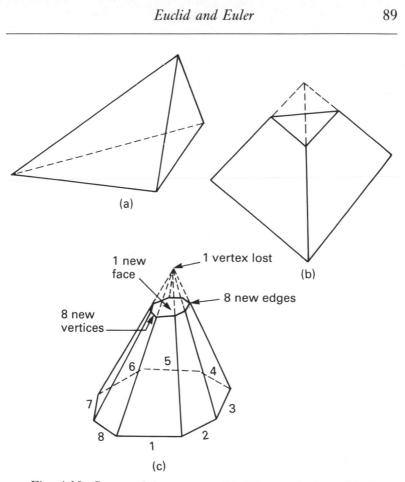

Fig. 4.10. Some solid geometry. (a) The tetrahedron. (b) A tetrahedron with one corner cut off. (c) A pyramid with an eight-sided base.

$$\begin{array}{ll} & \text{3 new edges} \\ & \text{3 new vertices} \\ & \text{1 new face} \\ \text{but lost} & \text{1 vertex} \end{array}$$

This is true of all plane-bounded solids (polyhedra) in which three edges proceed from each corner. A cube is another example. It should also be noted that the sides need not all be equal, nor the faces in any way regular, in order that this number of changes take place when a corner is sliced off.

Next, consider pyramids each based on a polygon; one is shown in Fig. 4.10c. If it has n sloping faces, cutting off the top corner will produce:

n new edges
n new vertices
1 new face
but will lose   1 vertex

It occurred to Euler that if the numbers of vertices and faces appeared on the same side of an equation which related edges, vertices and faces, the one new face would always compensate for the one lost vertex. Moreover, if vertices and edges were on opposite sides of the equation, then increases would always be in proportion, and this second requirement in no way conflicts with the first. So, starting with a tetrahedron:

(No. of faces) + (No. of vertices) = (No. of edges) + 2
      (4)   +   (4)   =   (6)   + 2

Starting with a cube, the equation reads:

6 faces   +   8 vertices   =   12 edges   + 2

and so on. The equation is the same for all polyhedra and is known as Euler's equation.

What is true for slicing off, of course, is true for adding on, and any plane-sided pyramid built on a face with n sides adds n edges, n faces and 1 vertex and loses 1 face.

Some solid geometry

We are more familiar with cubes than with any other plane-faced solid. Perhaps the sugar lump was responsible? I think not. We have always found boxes (if not cubical, then with a brick-like rectangular form) the easiest and most economical to stack. Our unfamiliarity with a cube is easily tested by asking the following question. If a cube be hung in space against a light background, so that only its outline or shadow is visible, and you are told that one corner of the cube is pointing directly *at* you, what shape do you see in outline? About seven people in every ten say at once: 'A square' and then add hastily: 'No, a diamond.' The correct answer is a hexagon, but it is less obvious that it is *regular*, i.e. all its sides are equal and all its angles are 120°. Look at Fig. 6.1c. Each side you see in outline is the result of a displacement along one of the three lines from the apex facing you and then a line *beyond* that. So each of the six occupies the same situation relative to you, the observer. The 'square' and 'diamond' answers seem to arise because the mention of 'cube' conjures up the thought that it is made up of *squares*. A square suggests four rather than three or six.

The second regular polyhedron is the tetrahedron (Fig. 4.11). It, too, has three edges running from each corner, and subconsciously one feels this to be more 'correct' than in the case of a cube, since all the faces are now triangles. It looks at first sight as if we could go on building as many regular polyhedra as we liked on the basis of (a) the number of sides of each face, (b) the number of edges meeting at each corner. But in fact it can be *proved* that there are only five such solids in all, thus:

(a) Shape of face	(b) No. of edges at a corner	Name of solid	No. of faces
	3	Tetrahedron	4
Triangle	4	Octahedron	8
	5	Icosahedron	20
Square	3	Cube	6
Pentagon	3	Dodeca-hedron	12

The three more unusual solids are illustrated in Fig. 4.12.

The five are often known as 'Platonic' solids, and for a fascinating reason. The Ancients, in their misguided search for Truth, decided that rather than look for an infinity of unobservable point masses by infinite subdivision, they could avoid the formless by dividing the seemingly irregular bodies which the world contains

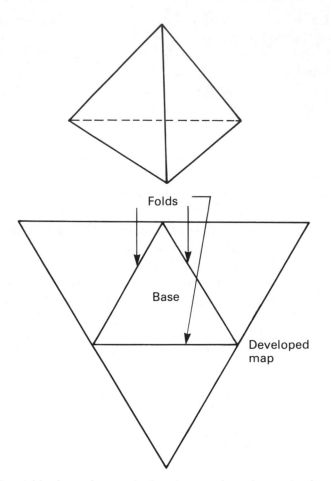

Folds

Base

Developed
map

Fig. 4.11. A regular tetrahedron has equilateral triangles for all
its faces.

into collections of *indivisible*, perfectly formed 'atoms'. It was Plato
who declared the four shapes to be four of the solids just listed,
each to be associated with one of the four 'elements' thus:

Cube	Octahedron	Tetrahedron	Icosahedron
(Earth)	(Air)	(Fire)	(Water)

The fifth solid, the dodecahedron, was thought to represent the
universe, its 12 faces being the 12 signs of the Zodiac. However,

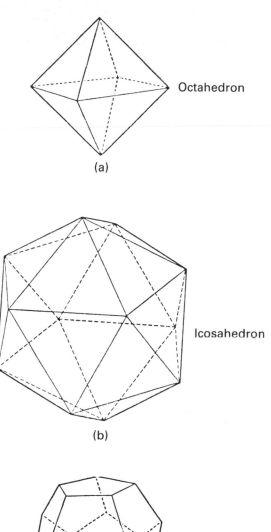

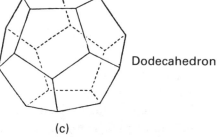

Fig. 4.12. (a) to (c) Three of the Platonic solids.

as D'Arcy Thompson points out, these five solids were known long
before Plato.

If concavity is allowed, as well as convex corners, then four more
regular solids can be formed by building pyramids on to the faces
of the five. These are known as the Kepler–Poinsot polyhedra and
are illustrated in Fig. 4.13. The two with star-faced (stellated)
dodecahedra were found by Kepler (1571–1630), the other two,
with star vertices, by Poinsot (1777–1859).

Martin Gardner points out:[9]

> Not so widely known is that there are an infinite number of
> semi-regular solids also with sides that are equilateral
> triangles. They are called 'deltahedra' because their faces
> resemble the Greek letter delta. Only eight deltahedra are
> convex: those with four, six, eight, 10, 12, 14, 16 and 20 faces.
> The missing 18-sided convex deltahedron is mysterious. One
> can almost prove it should exist and it is not so easy to show
> why it cannot.

It is almost certain that if an alien visited earth and tried to decide
which things were *natural* and which made by the hands of humans,
the only thing it would get wrong would be crystals, which, although
never *alive*, are both natural and Euclidean. But of course, our alien
must not be allowed a microscope!

What *cannot* be, in terms of regular solids (or their skeletons,
which are usually more important in nature than whether each face
is flat) is more common than what *can* be. We open the next section
with an example of the former.

Corals, Radiolaria and others

A common natural phenomenon is the deposition of inorganic
material (usually calcium salts or silicon) into or around a living
structure. The process leads to the formation of skeletons in such
far-developed animals as the mammals. It is the basis of coral and
of sea shells. But it can be traced back in size to sponges, Radiolaria
and diatoms. To set the scene and the place of Radiolaria in the

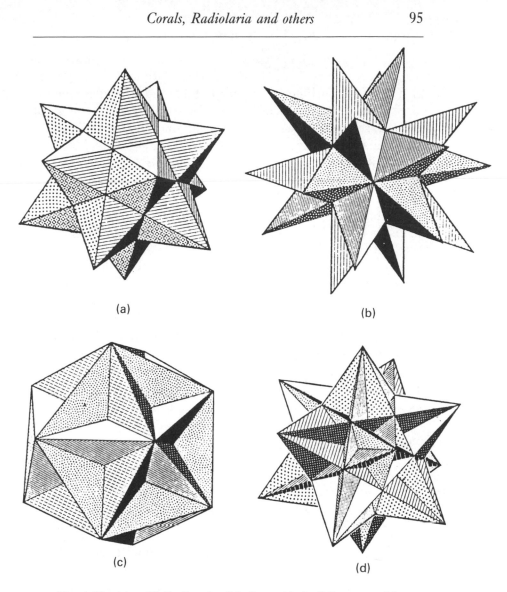

Fig. 4.13. (a) to (d) Stellated solids formed by building pyramids on or by excavating the surfaces of Platonic solids.

natural world, some dictionary definitions are almost unavoidable.

Protozoa make up a 'subkingdom' of the animal kingdom which is made up of non-cellular animals (i.e. not amoeba and the like) in which no nucleus is ever in sole control of any specific part of the living tissue. One class of Protozoa which consists of free-living creatures is known as *Sarcodina* and one Order of Sarcodina is Radiolaria. They are creatures that send out radial pseudolimbs, which, whilst not having the full capabilities of the limbs of animals, can be used for propulsion and for adhesion to surfaces. They are most certainly creatures of a *miniature* Lilliput. The surfaces of Radiolaria often appear to be covered with a network made entirely of hexagons. But we know this to be impossible, and if we look closely we can soon find five- and seven-sided structures also.

The finest artwork ever compiled on the structure of the Radiolaria is undoubtedly that of the German biologist Ernst Haeckel. A page from his book, *Art Forms in Nature*,[10] is reproduced here as Fig. 4.14. Top left, the structure is spherical but the six projections suggest an octahedron.* Top right, the centre is a genuine octahedron. The centre structure of the five is a regular icosahedron. Bottom right there is undoubtedly a dodecahedron.

In 1974, Dover Publications republished the plates from Haeckel's *Kunstformen der Natur* of 1904.[11] This is a publication that I commend to all readers. The illustrations might suggest that the artist had idealised the structures 'for art's sake', but Fig. 4.15 shows a photomicrograph of similar creatures, rather misnamed I always feel 'diatoms', for there are rather more than two atoms in each of these creatures (a few million million more!).

Beyond the Radiolaria

The development of the scanning electron microscope extended the size range that we could see and photograph enormously. The reason for 'scanning' is not obvious. When a TV camera 'sees' a scene, it does so as a human would read the page of a book, by starting at the top left hand corner and progressing horizontally,

---

* There is an interesting point here concerning the symmetry of points on the surface of a sphere. Neither three nor five nor seven points can be located so that each is related to each of the others in an identical way. Only the vertices of the Platonic solids can do this, i.e. four, six, eight, 12 and 20.

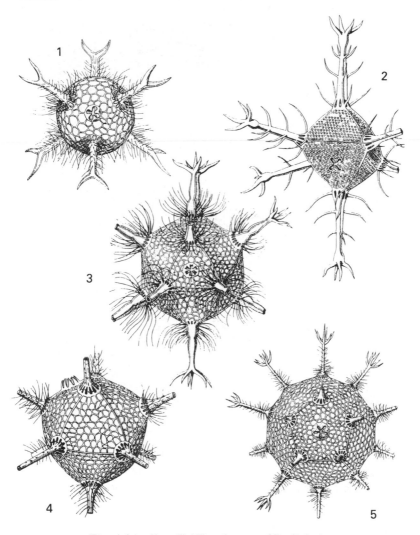

Fig. 4.14. (1 to 5) The shapes of Radiolaria.

line by line, down to the bottom right corner. Then it starts again,* and re-reads the page of 625 lines every $\frac{1}{25}$th of a second so that

* In fact, to improve the apparent continuity of vision the system is arranged to 'read' lines 1, 3, 5, 7 . . . in order down to 625. The it returns and reads the even numbers, 2, 4, 6 etc. We should get a funny-looking book if it were so printed! (Quite possible, but with no benefits.)

Fig. 4.15. Diatoms.

the eye can retain the image long enough to give the impression of a continuous picture.

Now, we arrange for it to do this because we cannot think of a better system at the price. But with a microscope it must surely be easier – you just look? To appreciate why this is not so, as the magnification increases we can think of light as a rain shower falling in straight lines on to the object viewed (Fig. 4.16). (The old corpuscular theory will serve us quite well in this respect.) In (a) four 'rays' of radiation are reflected to the lens system, and this is a measure of the brightness of the image we see. But an object only half as big, as in (b), only reflects two rays and is therefore harder

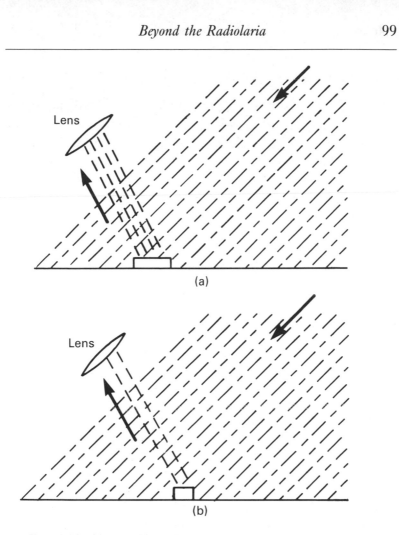

Fig. 4.16. (a) and (b) Radiation expressed in terms of a rain shower.

to see. If we wish to go on increasing the magnification, there are going to be two new requirements:

(a) we shall need more intense light on to the object (i.e. it has to 'rain' harder),

(b) we shall reach a size at which only one 'ray' is reflected, i.e. the object to be viewed is the same size as the 'diameter' of a ray of light! (size of a raindrop).

Now, an electron is of much smaller diameter and can be used as 'raindrops' in an entirely different radiation system. (In the early 1920s no good physicist would have accepted the concept that a stream of what were obviously *real* particles could act as a pure radiation wave in 'nothing'!) Thus problem (b) is overcome, but problem (a) is intensified. The new radiation has to be so powerful that it destroys the object we are attempting to see. Literally, it *burns* it up. But if, instead, we radiate only 1/10 000th of the area of the object at a time, and keep shifting the spot we are radiating as in a TV scan, any one spot can have 10 000 times the instantaneous radiation that does not quite destroy it for only a tiny interval of a time, and then have 9 999 times that length of time to cool down until it is 'hit' again.

There is, of course, one minor problem in that the human eye cannot *see* electron beams, but the screen of a cathode ray tube can, so our relatively long-known TV camera was at hand to do this and let us see the new world a decem away. We can photograph it too, of course, and in such a world, common objects look *very* different. Figures 4.17 to 4.19 show a range of such objects, from the surface of a flower petal to the antenna of an aphid and the surface of an eggshell. Now biologists and chemists worry about whether very complicated molecules are to be classified as 'living' organisms or just chemistry. Some viruses are *known* to crystallise into such 'macromolecules', and some of these, the measles virus, that which causes herpes, and others, make up regular icosahedra.

It has been estimated that there are of the order of 10^{33} living creatures on earth and that of these, over 75% are viruses. If this fact is inclined to make us 'feel small', remember that 250 000 000 000 000 000 000 000 000 000 000 is still a lot of creatures that are **not** viruses.

Nevertheless, there is the same amazing range of shapes and sizes in bacteria that we find in larger animals. For example, there is a marine bacterium called *Leucothrix mucor* that reproduces by tying itself (it is a long filament in shape) into a variety of shapes of knot that tighten more and more until they pinch themselves into two or more pieces. Be assured, nature really has tried every trick in *our* book! Of course, there had to be viruses that were helical. One such creature is that which causes mumps in humans.

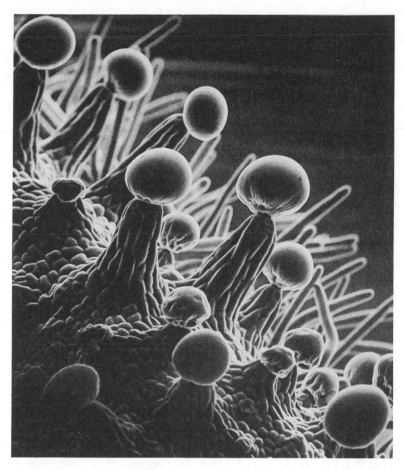

Fig. 4.17. Resin modules on the petal of a Marijuana flower
(×518).

It is interesting that in 1946 the state of electron microscopy was
such that it was not *quite* possible to 'see' individual atoms. It was
in an effort to increase the resolving power by just one more step
to make this possible that the late Professor Dennis Gabor invented
holography. That invention was to wait 10 years for the invention
of the LASER* to produce a strong source of coherent radiation,

* Light Amplification by Stimulated Emission of Radiation.

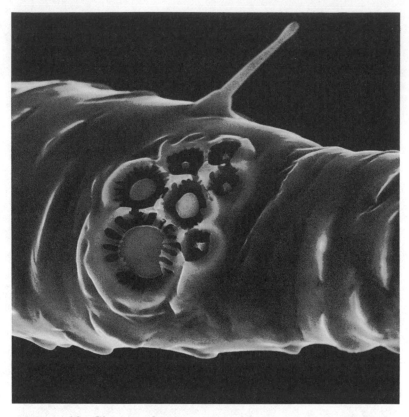

Fig. 4.18. Clusters of sensors on an aphid's antenna (×1862).

after which it had a dozen uses, none of which was improvement of the electron microscope. Science and engineering have a habit of happening just like that. (See also Chapter 8 on the subject of holography.)

Now that last step has been made and the electron microscope can 'see' the largest atoms – uranium and even down to atoms of barium. The smallest object now capable of being photographed measures 1/10 000 000 of a millimetre (1/250 000 000 of an inch), which is almost exactly a decem from the length of a human arm. Another useful exercise for trying to grasp the immensities of size range in terms of decems is that it takes light almost exactly 500 seconds to reach earth from the sun (to travel 93 000 000 miles

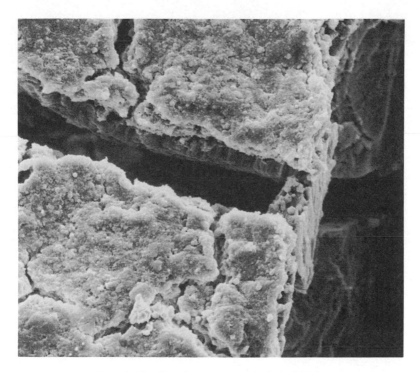

Fig. 4.19. Crack in an eggshell (×9811).

(150 000 000 km) at 180 000 miles/second or 290 000 km/second). So a 'light year' is $365 \times 24 \times 3600/500$ times the distance from earth to sun, i.e. just over 63 000 times, which is nearly 600×10^{10} miles (960×10^{10} km).

So we can span the range of human experience in these terms. From a uranium atom, magnified to measure 1 cm (0.4 inch) in diameter by the electron microscope, to the length of a human arm is a decem. From the length of a human arm to the distance between earth and sun is 15×10^{10}, which is just over a decem. (It is not '15 decems' for a decem measures $\frac{1}{10}$ of the logarithm of the ratio to base 10, in which case the exact value of 15×10^{10} in decems is 1.176 decems.) From the distance between earth and sun to the distance between the sun and the centre of our galaxy, the Milky Way, is just over yet another decem (1.023 D, in fact). From this distance to that of the furthest detectable body in the universe (catalogued as QSO OQ 172)* is just over another half decem (0.576 D).

Beyond the 'ends' of our journey through size there is neither 'nothing' nor anything we can begin to understand. But wouldn't it be *awful* if there were?

---

* And apparently travelling away from us at 0.94 times the velocity of light.

5

Our assessment of ourselves

Merely to dabble in philosophy is to come rapidly to the question: 'What am I?' The well-known answer: 'I think, therefore I am' is not very satisfying, but nevertheless the answer we have to settle for. In science, one can do little better than refer to a monumental statement by St Anselm who lived in the eleventh century and became Archbishop of Canterbury. After a lifetime of contemplation, he emerged with the words '*Credo ut intellegam*' – I believe, in order that I may understand. As far as we can see, and despite the rapidity with which discovery follows discovery today, it appears as if science may never have a deeper understanding than this. Define – or, what is a more honest word – *believe* in several 'fundamentals' (whether they be the earth, fire, air and water of the alchemists, the protons, neutrons, electrons and positrons of the 1930s, or the quarks, gluons, and whatever of last week) and the whole of science lies before you. Deduction follows deduction in a logical manner and all appears to fit together like a beautiful jigsaw puzzle in which we *know* that all the pieces must ultimately fit because it was originally cut from a single sheet.

But in science it is not necessarily so. The pieces fit nicely except at the 'edges' – the frontiers of knowledge where hypothesis follows hypothesis as each of its predecessors is destroyed by perhaps merely a single accepted fact that violates it. Logic? There are those who contend that the Devil put logic into the mind of man to confuse him, much as in the Tower of Babel relating to language – I must confess that there are times when I get just a whiff of this, but it is usually gone by tomorrow!

The fact that logic appears to produce 'correct answers' day after day, in millions of millions of different situations, is perhaps the ultimate miracle. Sir William Bragg wrote brilliantly on this theme under the title 'Concerning the Nature of Things',[1] a phrase that I have borrowed from time to time in this work whenever 'explanation' in its literal sense goes absent. I often think that we do not teach our schoolchildren to marvel at the right things. They crave science fiction and marvel at facts such as the laser (light cutting metal!), the electron microscope and its revelations, and space exploration. Would they not be better served to be reminded constantly that the law of gravity is merely the implication that because all unconnected objects have always been observed to fall towards the earth, those that have the opportunity to do so at any time in the future will always avail themselves of it.* The laws of physics are therefore no more than a collection of known facts, classified and labelled as a biologist might treat an order of insects. The possibility of the overthrow of a 'sacred' law by a single instance of violation should be constantly in the minds of the young or our progress rate must surely diminish.

In mathematics the situation is more clearly defined. You need to believe in only *one* thing and the rest follows. Define what is meant by the number *one* and you are ready to begin. But when you search for that definition, you meet the same barrier as that with which I began this chapter, for if you were to be confronted by an alien being who asked you, 'What is *one*?' I doubt whether you could better the reply: '*I*, am one' which simply takes you back to: 'What am I?' Whatever else we are we must also learn to live with the philosophy that leads us to say to all our fellow creatures: 'I cannot prove that *you* exist. All I can ever know is that *I* exist.' Taken at a lower level, this amounts to the frustrating thoughts, first that you can never explain what *green* is to a person who has been blind from birth, and second that among those

---

* Objects such as helicopters are 'connected' to the earth through the fluid, air. If radiation pressure is cited as a levitation method or if an inertial propulsion unit is ever developed, then the 'connection' would be through the ether.

who *see* colours, no-one can prove that, just because two people call rose-hips 'red', they each have the same experience when they *look* at red.

Such heart-searching has been going on over the centuries, often with little progress, so that we can marvel at the thinking of the ancient philosophers, but then, had they not the advantage of being uncluttered by science?

Defining 'the middle'

One outstanding facet of human life, as recorded faithfully by historians, is that we have always seen ourselves as the centre of attention, whether it be as 'I' in particular or the species as a whole. If two humans, A and B, could be put into space suits, tied together by a 10-foot piece of rope, transported to deep, inter-galactic space and set into rotation about an axis through the middle of the rope like a binary star, A will say that he knows he is rotating about an axis through himself and that B is moving around him at a radius of 10 feet. Evidence? A can see all the stars in the heavens passing through his field of vision and B blocks his view of the bit immediately in front of him. But B insists that it is *he* who is at the centre of rotation. Both agree that there is rotation, not only because they can see the relative motion to the stars but because the rope is in tension (centrifugal force effect).*

It was for precisely this viewpoint that Galileo's life was threatened because he would not declare the earth to be the centre of the universe, and he was made to recant. When it comes to

---

* A little problem to occupy the odd bit of spare time. In the situation here described where the gravitational force between two masses (each of 150 lb) is negligibly small and centrifugal force is clearly an *inertial* effect only, if the whole of the matter of the universe, except our two observers, their space-suits and the rope, could be removed, would the rope still be in tension? This question in effect says: the humans could not tell whether they were rotating or not, since they would have no frame of reference. But could the rope tell? This is a confrontation between the animate and the inanimate as profound as anything in the rest of this book.

beliefs, and especially religious beliefs, humans assume their most arrogant role. First of all, in biology they confer upon themselves the name *Homo sapiens* – Man the *wise*! What is wisdom if it contains no humility?

If scientific and technological progress over the next hundred years is only as rapid as during the last hundred (and there is every indication that it will be more rapid than more rapid!), we might have learned how to become invisible to each other as well as to alien beings and, if other worlds are watching us, as many of us believe, then it seems more probable that they would do it from an invisible 'hide' than from a landed flying saucer. This book is not concerned with UFOs or even with mystics or religions. This bit of science fiction is but another example of us seeing ourselves as the hub of all good things, the seat of God-like qualities, the centre of attention for alien beings. We should note that in holy wars, both sides are convinced 'God is with them', that they alone have 'right' on their side.

There is one further facet of human experience in which we see ourselves 'in the middle'. Any civilisation of the only kind that *we* know must come very rapidly to the questions: 'What is the smallest particle of "stuff" (scientific name "matter") that exists?' and, 'What is beyond the stars?' Piece by piece, limited always by our paucity of senses (most of us will only admit to five) and by our technology, we put together a model of the structure of matter. At the same time as we were making worlds in miniature we were mapping out the shape of the solar system in three dimensions and becoming alarmed by the similarity of the two structures. They are sufficiently similar to suggest that there might be equations common to the two that are basically responsible for the similarity. Often we come close to saying that there is an attainable entirety of knowledge, not now far beyond our reach. That is what the old alchemists believed, too. Similarly there are those today who seek the 'origin of the universe', not paying too much attention to the fact that to do so is to assume that it *had* one – a fact that is questionable for those who dare to think.

I am sure that I have covered enough facets for readers to appreciate how self-centred are our ideas about ourselves, right up to the point where we see ourselves 'in the middle' as regards size – just about 'half way' between a neutrino (than which there is not a thing a great deal smaller) and a spiral nebula (than which few things are bigger). In fact we are so much 'in the middle' by what might even be called a commonsense view of things that it is surely time to ask whether or not this whole concept of a middle is a dreadful illusion, of the same nature as the consideration of Oxford Circus as the centre of the world's surface (as has been said, albeit with tongue in cheek), yet one may proceed ever westward therefrom and in due course arrive at Oxford Circus again from time to time! If 'space is curved in the presence of matter', a deduction made from Einstein's views of relativity, then perhaps so is time, his so-called 'fourth dimension' – and what of Ouspensky's three dimensions of time, as there are three of space?

No, I refuse to take my 'amateur' philosophies any further. I am merely moving towards what we must all surely agree in regard to our self-assessment. We are an arrogant, emotional, ignorant, near-megalomaniacal race of creature.

The human communicator

If the insects of the world could hold a conference on the state of evolution and the special place occupied by the mammal *Homo sapiens*, they might well reach the following conclusions in relation to the latter's awareness of the universe.

> Sight is easily its best sense. It is not so good at seeing as are our arch enemies the birds but it has extended its range of perception both in the direction of the microscopic and the macroscopic by the use of sophisticated tools. Its sense of hearing is a whole order of magnitude worse than that of mammals, birds and insects. Its sense of taste is distorted; its sense of smell is almost dead. Its sense of touch it forgot it ever had, and the senses that are most important of all, it does not believe in!

All that, of course, is going to need defending. Let us examine each item in turn. Our eyes are good, not only because we have choice of rapid re-focus and re-orient, nor because we have a wide field of view, but because the information about what we see is conveyed to the brain in *parallel* paths. It solves a basic communication problem. If there were only one telephone line between two major cities and the telephone was so primitive that it could only relay one conversation at a time, it would be a virtually useless system. It would be an example of a *series* system at its worst. In electrical engineering, the word 'series' has a special meaning in that it implies that a certain 'flow' quantity passes through all units in a line. 'Parallel' in the same context means that a plurality of units side by side are all connected to the same pressure, and each may carry different flow quantities at the same time.

But the words 'series' and 'serial' are related, and when we speak of a series system in the context of *time* we mean that each bit of information must have its own bit of time allotted and that the bits must proceed along the communication link in order, like an orderly queue at the baker's when there is a bread strike – and we are well aware of the penalties of such a system. What is more, we even know the penalty of using a time series in systems that work incredibly more rapidly than does the human brain – computers and TV sets.

Sight

If you remember, we were about to consider eye–brain communication when I digressed into an elaboration of the concept of series and parallel (as used in the context of electrical circuits) in a time sense. Each of our eyes has over 150 000 parallel channels of communication with the brain, and the message of what the eye sees is communicated in *parallel*. Contrast this with our, surely now most beloved (even more than the motor car), invention – TV. The TV camera 'scans' the picture it is focused to communicate by looking at a tiny area of the picture. The area inspected shifts in an orderly fashion along horizontal lines (625 of them in the UK) so that the whole of the picture is scanned in $\frac{1}{25}$ of a second. It exploits the natural phenomenon that we call 'persistence of vision' whereby we accept two different pictures that are flashed to us less than $\frac{1}{25}$ of a second part as the *same* picture, and if small differences are included in each successive picture the eye/brain combination interprets the whole as *movement*. Hence we can have the 'movies', TV and other illusions (for this after all, is what they are). This is not to say that if successive pictures are sufficiently different, we cannot detect anything but a blur. To test this, move your hands, with fingers separated, rapidly from side to side under the illumination of a fluorescent lamp (as opposed to a filament lamp which never really 'goes out' between one peak of alternating current and the next because the filament has thermal capacity (it stays hot) and does not have time to cool down very much in $\frac{1}{100}$ of a second). You will see your hand as a series of images. The retention of vision enables you to see many images at a time, but the speed of response is such that the images are as sharp as the shortness of the light pulses allow. But the TV scanning camera technique converts a 'picture' (a phenomenon of parallel communication that we have always taken for granted) into a serial train of brightness and darkness signals so that the transmitted signal frequency must be extremely high if we are to have all the information from one picture in $\frac{1}{25}$ of a second. If the picture clarity is to be as good in vertical columns as it is in horizontal lines, of which there are 625, then 625×625 bits of information occupy $\frac{1}{25}$ of a second, and if the picture were a pattern of 625×625 chequered squares of alternate black and white, the frequency to be communicated is $\frac{1}{2}(625 \times 625 \times 25) = 4 \times 10^6$ Hertz (Hz). Now, radio transmission demands that such information be sent as variations of a wave of even higher frequency (a 'modulated carrier' wave, we call it). The modulation can be in amplitude or in frequency or in phase, but in all cases the frequency of the carrier must be at the very least 10 times that of the signal, so our TV transmissions are at Very High Frequency (VHF) and we run short of 'bandwidth', one of the results of which is that TV signals can only be properly communicated over virtually visual distances.

So a network of land lines connecting many transmitting aerials is necessary so that each viewer's private aerial can 'see' at least one transmitter. Until space satellites were developed, TV across the world was not possible by radio waves. The lower frequency radio waves can be 'bounced' off the ionised layers of the upper atmosphere, and this was exploited before it was understood (as is the case with most good engineering inventions). In other words, we used radio to investigate the upper atmosphere rather than we designed radio to use what was known to exist. Artificial satellites have done for VHF what the ionosphere did for sound radio at lower frequencies.

Therefore, we can see first that TV is a whole degree less 'clever' than the eye, and second that it is essentially a *wasteful* system. When we watch a play, on TV, or we are shown round an art gallery or watch the news, notice how much of the picture is actually changing from fraction of a second to fraction of a second. The answer is: 'very little'. It may only be a person's eyes, or mouth, or in the case of a work of art, nothing for several seconds. Now this ought to be exploitable. It was exploited in the cinema to tremendous effect by Walt Disney and his kind. In cartoons, the artist need only alter the moving bit, from frame to frame, cutting down the amount of work needed to draw each picture by a factor between 1000 and 1 000 000, depending on the action. But we have not so far exploited this feature in TV at all!

But let us now ask whether or not nature has made use of this, and the answer hits us squarely in the face – of course! Try looking for a tiny bird or insect among foliage. It will be invisible – unless it moves! Then you see it at once. Moving things are far more 'obvious' than are stationary objects to most creatures that have eyes, not merely *H. sapiens*. Otherwise why would so many creatures 'freeze' on the suggestion of an enemy's approach? Many creatures move rapidly, with long periods of 'frozen' position between movements when there is no obvious danger. Large animals, through squirrels and rabbits to hoverflies, pond skaters, and many other tiny insects right through the living world, practise this technique. Not only the victims but the predators also must avoid movement as far as possible. In my young days I spent a happy hour sitting high up in a tree watching the antics of parent and baby rabbits at play. When it was tea-time I decided to make a sudden noise and watch them dash for their burrows – I coughed. Then I said loudly, 'I'm going home now.' To my surprise all the rabbits froze, with ears erect. None moved a muscle. I clapped my hands. If they moved at all it was a twitch, no more. Finally I imitated a dog's bark and apart from minute ear twitches, nothing moved. Then I moved one of my arms covered in a white shirt sleeve. In seconds all rabbits were safely in their burrows! Sight is the ultimate criterion of danger prediction for many species of creature, but it is inextricably tied to the concept of the 'bit' that moves in an otherwise static background.*

So our TV camera is a poor substitute for the eye. It is even thought that the human eye can detect a single photon – the smallest quantity of light to exist – but our acuity of vision is, as the Insect Conference concluded, below that of the owls or the hawks. The range of frequency that we can see is quite small in comparison with what is, so to speak, 'available'. The whole range of electromagnetic waves extends from one cycle per second (1 Hz) or less to 10^{25} Hz or more. In music an octave is defined as the space between two notes, one of which has double the frequency of the other. In this terminology our visible range covers almost exactly one octave – the frequency of violet light being almost exactly twice that of red light. Radio waves of lower frequency could be said to occupy 37 octaves, merging at the higher end into infra-red, of which there are some 12 octaves. Beyond the violet end of the visible there are five octaves of ultra-violet, followed by 14 octaves of X-rays; higher still are at least 11 octaves of gamma-rays, and beyond this? – we are still looking! But it would be quite wrong to assume that other 'seeing' species were restricted to the same boundaries as are we. Bees are known to be capable of vision, as we understand the word, well into the ultra-violet. My own researches on moths suggest

* In communication theory it is well known that a signal (audible or electrical) can be detected against a background of 'noise' in excess of the amplitude of the signal (signal : noise ratio < 1) if the signal be modulated. See also pp. 198–99 on the subject of moth communication by infra-red.

that their vision extends to the infra-red. Dogs are said to have only 'black and white' vision and I myself once knew a man so colourblind that his view of the world was virtually just that. He could not distinguish between the green, brown and red snooker balls on a billiard table, bright as they are. To experience his world, all we have to do is to turn down the colour on a TV set during a snooker match. Yet we need not commiserate with dog or colourblind man, for all of us were quite satisfied with 'black and white' TV until we had something better. *All* things are relative, not just Einstein's concept of velocity.

What is more intriguing is that I also know a man who suffered terrible spinal injuries and after years of drug treatment suffered experiences that in many ways were worse than those resulting from *no* treatment (including hallucinations). After he recovered he told me quite cheerfully: 'Do you know there are colours you have never seen?'. I nearly made an ass of myself by exclaiming, 'What are they like?'. Just in time I realised how meaningless that question would have been. But my horizons broadened a little and my humility increased a little on that day.

We make photographic emulsions that are sensitive to infra-red and ultra-violet, and most films for ordinary daylight photography will record X-rays, gamma-rays and beyond as 'white'. They can never help us see what the bee or the moth sees. We should never assume that a green aphid that appears to us to be well-camouflaged on a green leaf is equally difficult for a beetle or a bird to spot. A photograph on an ultra-violet or infra-red film may show it as a dark insect on a light leaf or vice versa. We must be ever watchful of assessing other creatures on the basis that they must be, of course, *like us*!

Sound and language

We often tend to think of vision as an experience more than as a means of communication. Yet if that were so, few books, and especially technical books, would be 'illustrated'. Sound we regard as 'communication proper' – the giving and receiving of messages. Even with music (from classical to punk rock), which might reasonably be described as a 'recreational experience', the enthusiastic performer and listener both claim that the former is communicating his or her feelings to the latter.

By sound we teach our children from the cradle, and thereafter by sound we teach our contemporaries. So far as we know so do mammals, birds and possibly fishes. The major difference is in the emphasis that is put on this form of communication. In our own case it is the very bedrock of education. Not satisfied with a spoken language, we developed a visual form – the printed word, which is so glorified that Nobel prizes are awarded to those who put their words in the most elegant order. We tend to forget that the alphabets of the world began as *illustrations* that degenerated in the cause of simplicity, unless of course we appreciate Chinese writing, which is pre-Roman, pre-Arabic, pre-most things.

Animals, of course, are *primitive* in their languages. We would not expect more of 'lower' creatures. Might we not, however, ask the question: 'Do they *need* a sonic language so much as do we?' The giraffe (the only truly *dumb* mammal) traded-in its voice for the ability to reach better food. Giraffes must have evolved something to replace sound for they would survive (in the absence of *H. sapiens*) as well as most species of mammal.

The problem of language began long before the Tower of Babel. Once we could speak we almost abandoned gesture, the dance and most of all the use of face muscles. The totally deaf person manages quite well to be understood, but suppose there had been a species, *Homo silens*, with no vocal chords, evolving for 10 million years. Might it not have rejected all but the simplest of sounds, knowing that its eyesight was a far more developed sense? Just suppose you were only allowed one word in a foreign language with which to make your way in the country of its origin. Which would you choose – 'Water', 'Stop!', 'Please', 'Good'? Then try being allowed two words. 'Yes' and 'No' appear to qualify, except that they are so readily replaceable by up and down, or left and right movement of the head, hand or foot. Many birds have at least *one* 'word' that is the equivalent of 'Achtung', whether or not they are German birds.

But because there are millions of species of living creature, we cannot dismiss them all as having a less complicated language than ours. For one thing, as with light, the range of sound waves that we can detect has an upper limit of 15 000 cycles/second,* when we are young, and recedes to 10 000 or below as we age. Dogs can hear at least up to 20 000, hence the 'silent' dog whistles that can be bought. I myself have developed a means of generating 'sound' between 15 000 and 20 000 cycles by drawing in air between my top lip and my lower front teeth with the two pressed tightly together. I have fun attracting the attention of dogs, who are taking their owners out for 'walkies', without the owners being aware of it. Insects use higher frequencies still, possibly up to 100 000 cycles/second or more, but some share a common bandwidth with their predators, the bats, for obvious reasons of defence. The moth/bat range is centred around 50 000 cycles/second.

The creatures that disturb me most are undoubtedly the dolphins. There are many stories of dolphins, some legend, some undoubtedly true. They appear to be unique in several aspects of life. For example, they are the only animal known to display compassion (by human definition of that word) to animals of another species – such as ourselves. They can be taught human speech as no dog can ever be. Our domestic pets respond to a combination of gesture, loudness, facial expression and familiarity of sound (without hearing the latter as 'words'). This statement will undoubtedly be challenged by dog-lovers of all shapes and sizes (the owners, not the dogs!), but let them try this experiment. The next time Bonzo comes in with muddy feet, instead of pointing to the door and shouting in anger, 'Get outside!', carry out the same gestures, facial expression and the same angry appearance but shout, 'Strawberry jam!' Bonzo will go out just the same! A wild dog seldom barks. It howls and growls. The bark is its attempt to imitate human speech. We are, it appears, very poor teachers.

Dolphins chatter away to each other whilst performing circus

---

* Middle C on a piano is 256 cycles/second. The C an octave above is 512, and so on, and most pianos run out at the top end at about 4096 cycles/second.

tricks for us in artificial pools. I can only describe one of the sounds they make as being the result of running a tape recorder backwards at high speed with the output then re-recorded and that second recording again run at high speed. Many humans have tried to make a 'language' out of what certainly sounds like the very highly compressed speech of the dolphin and all have failed. Possibly the sounds we hear are only bits of the sentences, and the missing milli-seconds, of what is to us silence, are filled with other 'words' at 30–60 thousand cycles/second. Or possibly the gaps are a code in themselves. Certainly it has been established that the songs of certain birds which appear to be broken up by short intervals of silence are in fact continuous songs, modulated in frequency with the 'silent' periods merely being ascents into the human supersonic zone.

Some say that dolphins came out of the sea millions of years ago and, finding the land such an inhospitable place to live, returned to the sea – and *that* disturbs me too, because it just might be true. They may well be more intelligent than us (measured on a scale different from our IQ tests). When they swim, they leave no turbulence behind them as do all our water-going machines and most, if not all, fishes. Could they be indeed the 'master race'?

The thing that is basically wrong with speech and language is that it is an essentially *series* affair. Every sentence in a story must be in its proper place. Every word in a well-written sentence must obey the rules of grammar. Every syllable in a word must be in time order, and even the train of sound waves that comes to the ears of the listener from the mouth of the speaker must be in a serial train of pressure waves, and the air between mouth and ear is a form of gaseous tape-recorder, no less. You must 'play' all of the tape to understand any of it, in general. Series things, as we saw with TV, take too long to transmit and receive which is why, dolphins apart, other creatures developed other methods for sophisticated communication, retaining sound only for emergencies when Achtung!, Help! and a few others would suffice.

It took Sir Walter Scott a whole chapter in *Ivanhoe* to describe Cedric the Saxon's banqueting hall. A single coloured picture would have communicated the same information in a matter of seconds. A holographic, three-dimensional manifestation would have been better than the book. We should never have departed from the cave men's drawings. Figure 5.1 says very quickly, in *any* language, 'At noon I killed a stag.' Picture communication is quicker than speech, provided the pictures can be reduced to the absolute essentials. But then, the written word can be shortened a good deal without loss of meaning provided one keeps within a narrow subject and titles it correctly thus:

HOUSES FOR SALE
B'pool ½ m sea ft. Det. Mod. 3 Rec.
Dng. Kit. o/s w.c. 4 Bed. Gar. Gd Gard.
Fhld. £34,000 o.n.o. Tel. 5128 Ev.

Fig. 5.1. Pre-historic cave drawings.

But the same basic series limitation remains. The fact is that it takes us 21 years of a child's life to bring it up to cope with the world, let alone reach the frontiers of knowledge. A fruit fly only has a few weeks of life altogether! Whilst it is true of the fruit fly that it can inherit nearly all the facilities it needs for survival and reproduction, it is less true of a social insect such as a hive bee, a common wasp or an ant. If ants used a serial language to instruct each other on the subject of work required, they would use a considerable fraction of their lives in performing only one simple task each. But of this topic more under the heading of 'Touch'.

My final thought on the handicap of language is best illustrated by an actual incident that occurred during an international conference at which a friend of mine was present. Various delegations each had their own interpreter in a glass box from whom they were receiving continuous translation as a delegate from another nation addressed the assembly. A German was holding forth with some vigour. A long period of silence from the UK interpreter caused my friend to glance at the glass box to see if he was unwell. No, he was quite well, but obviously very tense. Another 20 seconds and he was gripping the stand of the microphone and his face was bright red. Still the German went on excitedly. Break point had to come. Through the headphones of the astonished UK delegates came a desperate roar: 'For God's sake, man, the **verb**!' No more fitting epitaph to serial speech and the written sentence is needed.

Taste and smell

Taste and smell are very closely connected, and many readers may well declare them inseparable. But differences there must be or no-one would ever eat Gorgonzola cheese or Durian fruit! Likewise, some pipe tobaccos are intriguingly fragrant to those around the smoker, but the latter knows that they taste less pleasant than other brands and burn hot. It is also necessary to point out that most scent experiences are not the result of inhaling a gas so much as a particulate smoke, where much bigger molecules are absorbed by the mucous membrane, after which a very able band of 'analytical chemists' gets to work identifying the substance, even though the latter is a mixture of long-chain molecules, as in the case of fried pork sausage.

That scent in humans is a relatively crude sense is not in doubt. We cannot follow each other across country from the smell of crushed grass where the feet have trod. The fact that the dog can do this even when the field has been spread with rich manure indicates that the dog's olfactory sense is selective, as is our own sense of hearing. By this I mean that when we attend an orchestral concert we can, if we wish, devote the whole evening to listening to the oboe, or, by 'throwing a switch', we can 'tune in' to the French horn, or we can enjoy the whole ensemble. It is purely a matter of choice. But we cannot distinguish between two bits of blotting paper, one of which carries a smear of bacon fat, the other a drop of apple juice, if both have been soaked in French perfume and are still wet. The fact that animals are scent-selective makes it a fair bet that the ability extends to taste also, and a dog's dinner must contain the same aesthetic qualities as a Beethoven symphony. The expression 'dressed up like a dog's dinner' suddenly assumes a new emphasis. Imagine being able to elect to taste the subtle flavour of black cherry jam with a mouthful of piccalilli!

It has been my experience that it is almost a legend that children like sweets above all else. Many very young babies, tasting their first semi-solids (soups, etc.) prefer a meat taste to a sweet one. They are born carnivores whatever our consciences lead us to believe in later life. Another tradition forced upon children is that of association between taste and appearance, especially colour. Blancmange is *pink* is it not? Who would prefer butter that was white? The answer to this is known, for it was tried commercially (most butters contain colouring matter). Only a very small percentage of customers bought the white, natural butter. Even white chocolate never quite caught on, although its taste is almost indistinguishable blindfold. We have distorted our sense of taste until things taste like they look, and if they do not, we are thrown into confusion, even near-panic for quite a time. I remember picking up a tiny biscuit at a very polite cocktail party, believing the biscuit to be sweet and decorated with sugar cream. The first reaction of my taste buds produced a mental 'My god, I'm poisoned!' reaction which was smothered by the civilised smile demanded of the occasion. It took me at least a whole minute to discover that the biscuit was savoury and its dressing was anchovy paste.

That our tastes are conditioned by the chef is to be seen in our acceptance of bacon crisps as bacon. We are so good at manufacturing potato crisps with a built-in flavour that we have effectively been reduced to eating what the accountant rather than the general practitioner says we should! *Homo sapiens* can no longer tell what is poisonous. Some of us can undoubtedly train the palate to specific tastes so that we can tell the year and place of origin of a wine. Why else would individual bottles be sold at auction for thousands of pounds each? Yes, it is a form of snobbery, but not without physical evidence to support it. But the expert wine tasters are no better than the rest of us in knowing what substances will kill them. In matters of taste we are as helpless as are day-old, blind, naked mice. We have to trust each other in accepting offered food. Such trust, of course, came with civilisation. We couldn't lose quite *all* the time.

Smell is a dying sense to some of us more than to others. (The charwoman in Jerome K. Jerome's *Three Men in a Boat*,[2] when the cheeses that had emptied a railway carriage and caused a horse to bolt were placed right under her nose, declared that perhaps she could detect a faint odour of melons!) Yet it is also on record that a human nose has detected a single molecule of a very smelly, highly volatile substance. This claim was based on an experiment in which a bottle of the substance was opened in the corner of a roomful of people who, without looking round, were to indicate when they detected the smell. Statistical methods were used to determine the speed with which the molecules diffused and the person in the diagonally-opposite corner to the one in which the bottle was opened was deemed to have detected the first molecule to arrive. This statement should be contrasted with another assertion that I have seen in print, that a bloodhound needs a minimum of 1500 molecules per cubic centimetre (cc) in order to track a source successfully.

One final point about smell. Unlike the sense of hearing, whether in humans or beasts, a sense of smell can give you no information about its point of origin. True, you can face the wind and declare the smell to be coming 'down-wind', but this is no indication that the air has come to you in a straight line. In still air you must always 'go looking' by moving first this way and then that and effectively seeing whether the number of molecules per cubic centimetre increases or decreases. So our attempts to detect a gas leak produce similar movements to those of a dog tracking a weasel. Memory, and the ability to dispose of particles received only a second ago, are vital in this exercise.

This limitation of smell to disclose the source is not found in acoustics because animals have *two* ears. The wavelength of sound in air (speed $\approx$ 1100 ft/second or 335 m/second) for a frequency of 1000 cycles per second, for example, is $1100/1000 = 1.1$ feet (33 cm). Two ears spaced perhaps 0.8 feet (24 cm) apart will register phase differences, from which it is clear that the information about phase, including lag/lead, can be used, as the result of *learning*, to indicate the invisible source of a sound.

Touch

Why might the insects, at their 'conference', have declared of our sense of touch that we 'forgot we ever had it'? The answer is not by any means obvious, so we shall need to examine ourselves very carefully in respect of this sense. When asked which part of us we use principally as tactile communicator we are surely agreed that it is the hands – the fingers, that daily tell us whether bread is stale, which key (in the dark) is the one to fit the lock, the quality of a fabric, the way to push a pen (we can write quite well in total darkness). Yet our fingers are so insensitive that most adults can cut quite thick slices of skin from their fingers without drawing blood or without feeling pain. Through work, our nerve endings have shrunk well below the surface for protection. Card sharpers are known to have sandpapered this hard top layer from the ends of their fingers in order to detect small differences in shape along the edges of playing cards. Yet a baby only a few months old knows better! Every new thing it encounters is 'examined' for a few seconds only; then, into the mouth it goes! Baby is neither teething nor hungry. It knows that inside its mouth it has a tactile receptor dozens of times more sensitive than are its fingers. So this bunch of keys 'looks like' *that*, thinks the baby. It feels like *that* (to the fingers), but what is it *really* like? It always lets the tongue have the last go, the most reliable assessment. Of course, as soon as proud and caring parents have taught it serial speech, any further attempts at accurate tactile measurement will be rebuked with: 'Take that thing out of your mouth, it's dirty!' Not until many years have passed and the young human is ready to make love and to re-create its kind, will the tongue be properly used again as communicator.

The tongue is more than a mere receptor of information. It is a tactile *amplifier*. When you have a hole in a tooth it feels like a huge cavity to the tongue. This facility was developed in us to prevent us from swallowing hard objects that might choke us. So effective is this amplification that many children have difficulty over-riding it so as to swallow a pill only the size of a pinhead. I know that *I* did as a child. My mother would put the pill in a spoonful of raspberry jam (seeds and all). I would swallow with exertion several times, really *trying* to swallow the lot, but the pill would remain! It was a sphere, after all, and a raspberry jam seed is not. Our tactile sense is virtually as accurate as our eyesight, if we were to allow it to be so. Dogs and many other animals do a great deal more licking than do we. Licking is sociably unacceptable. What I found encouraging, as a boy, was that my dog was no better at swallowing pills than was I. A doggie pill was inserted into a piece of raw meat. The dog would swallow the meat almost at once, then the tongue would flick out the pill – but, of course, the dog had an advantage beyond the mere shape and hardness. It could taste the pill selectively.

Perhaps the only human example of genuine tactile communication that contains an element of 'parallelness' is when two old friends meet after years apart, and greet each other with the traditional handshake. There is an exchange of sentiments in such a situation that can never be put into words.

The other senses

What then remains to discuss of our communication senses? Very little, if you are an orthodox scientist who regards extrasensory perception and altered states of consciousness as not even 'respectable'. What evidence have I then, that there is more? I have seen flocks of starlings 10 000 strong, turn a right-angle in unison without an obvious leader. I have read that the only mammal that would die of thirst, if placed a thousand times its own length from a water hole in a desert, is *H. sapiens*. The others would *know* it was there. It will not suffice to credit them with the ability to smell water. It will not suffice to multiply the sensitivity of our own senses by a hundred and credit the lower animals with this ability. As we saw with the bloodhound, the important feature was not the molecules/cubic centimetre sensitivity but the selective ability.

Our assessment of ourselves as communicators is grossly overrated. Let us test our viewpoint in other facets of life.

Humans as technologists

Comparatively recently (in 1978), I saw an article that stated emphatically that nature was not 'good enough' to have discovered pure metals and had to 'make do' with other materials. I hold quite the opposite view and am in total agreement with my ex-colleague at Imperial College, Gilbert Walton, who has written splendidly on this and other aspects of animate and inanimate things. He says in one of his articles:[3] 'As far as is known complex natural things have never been made of metals. No doubt if we knew how to do so, we would abandon metals in favour of organic materials which have a far greater variety of properties.' The same is true to a lesser degree of the wheel. Nature laughs at our 'iron birds' and 'iron horses' and doubtless wonders how long they can survive. The coal and oil that they eat were once alive. How long before we learn that living things were born to eat other living things and that death is the only means to progress? We have made a start. Plastics have revolutionised the modern kitchen. Yet within a generation we realised that plastic materials were themselves a form of pollutant. An empty polythene carton tossed carelessly on to a rubbish tip is there for 10 000 years. Work began on plastics with a life span, plastics that would be changed chemically by the ultra-violet that gives us a sun-tan in summer. The materials of our tools, our luxuries, our sports must be re-cycled like the materials of our bodies. 'A man dies,' said Leonardo da Vinci, 'always to be re-born in part.' Gilbert Walton elaborated on the re-cycling of materials and the apparent wastefulness of nature that is *only* apparent.

> The basic molecular material, called nucleic acid or 'DNA' does not appear to vary much from one organism to another. Combined in the nucleic acid is a limited number of amino acids, and the order in which they are linked appears to carry the complexity. The situation seems to be closely comparable to our own writing and printing which is used in the recipes for making things, whether it be in a cookery book for making a pudding, or in the construction manuals for elaborate machinery. Here ink and paper are the basic materials which only vary in unimportant ways from one document to another. There are comparatively few letters in the alphabet, but the arrangement of these in sentences can carry an infinity of different instructions . . .
>
> The comparison between printing and genetic material may be carried further. It is often pointed out that nature is exceedingly wasteful. It is worth noticing, however, that the waste occurs in seeds and spores and eggs, and not in bulk organic and nitrogenous material which is carefully conserved in elaborate cyclic processes. Of all the acorns from an oak tree, very few will grow into new trees, and of all the eggs from a fish few will hatch. It is a strange point that we ourselves waste paper and printing on a prodigious scale. We have only to think of the daily newspapers, the technical and trade journals, the advertising, let alone spoken words and talk . . .
>
> In the origins of our language people must have appreciated

the comparison between writing and seeds because the same
words are used for both. Literature is disseminated, books
are conceived, ideas germinate, plans hatch and sermons are
delivered. The parable of the sower has a double meaning;
'the seed is the word of God'.

So it appears that nature adopts the slow, series communication
for the passage of the characteristics of one generation to the
next, a process that is itself a slow one. But for urgent communications it spurns the printed or the spoken word as it does the
wheel for locomotion or metals for the bodywork of creatures
and for their limbs (tools).

Homo sapiens, biologist

One might distinguish us from all other living creatures by saying
simply that all the others 'play according to the rules'. We go out
of our way to break them. This is by no means a condemnation
of humans. It is the manner of the breaking that is vital. As a
race we do not, of course, even conform with each other. A bird
will fight another bird of the same species for 'territory', but rarely
to the death. Desmond Morris, author of *The Naked Ape*,[4] points
out that neighbouring primitive tribes in New Guinea also go to
war for territory, but an important secondary reason is for individual men to get wounded and to return to show their wounds
to their womenfolk so that they may be seen to have sustained
them while attacking, i.e. on the fronts of their bodies.* But the
New Guinea 'savages' stop the fighting if someone gets killed,
and both sides mourn. They take no shields into battle, nor do
they tip their spears with poison, even though both techniques
are known to them. They are, like the animals, 'playing to the

---

* Shakespeare records that the same view was held in ancient Scotland
(*Macbeth*, Act V, Scene VIII):

   Had he his hurts before?
   Ay, on the front.
   Why, then, God's soldier be he!

rules of life'. One of the few shining lights in the modern human ways of life is the fact that we managed two global wars without the use of germ warfare even though the technique was known in the first and such weapons prepared for the second. Likewise, we fought out the second without the use of gases, which techniques had been developed to a far greater degree of horror than was known in the first, bad as that was.

We broke the rules in making wheels. We broke them in refining metals. But we also broke them in healing the sick, in caring for the malformed and the weak, in making childbirth a whole order of magnitude safer. We broke the rules in selecting edible vegetables and protecting their growth until their population was thousands upon thousands of times what was intended. Many authors have condemned such desirable activities as building airfields, irrigating deserts and developing insecticides and weedkillers as 'disturbing the balance of nature'. But each of these and many other aspects of civilisation disturb it far less than does the showing of compassion to the starving, the fight against fatal diseases and all the medical research that prolongs the life span.

An advertising brochure issued by the Plant Protection Laboratories of Imperial Chemical Industries includes a contrived interview of one of their staff by an over-critical journalist (which latter subspecies is unduly common!), who asks, 'But aren't you afraid of disturbing the balance of nature?' The reply is a message for all mankind, so that he may know just how many sacrifices have been made, that he might occupy the place that he does on earth: 'There is no such thing as the balance of nature. There is nothing so unnatural as a field of wheat.'

The one thing that those laboratories find most difficult to do is to grow 'standard weeds'. A hundred seeds of groundsel, all as like as the proverbial 'peas in a pod', will produce seedlings of a variety of sizes. Select ten of these that *are* the same size and shape and subject them to identical conditions and some will grow faster than others. Each generation of weeds has its giants and its runts, and to a vital purpose. A good many lawn weedkillers rely on the principle that thin grasses will shake off the chemical whilst broad-leaved weeds retain it until it kills them. Such weedkillers destroy the giants, the ordinary and the small, but the runts survive, like the grass, and for the same reason. Yet the seeds from a runt are as capable of producing offspring of all sizes, including giants, as are those of average size. It is as if weeds anticipated the ascent of man and his inevitable lawns. Comes a drought, however, and the runts are destroyed. The giants' roots are big and grow deep, and though the leaves are shrivelled, the roots survive and produce new growth when rain finally comes.

Little wonder that farmers are ever worried about their crops. They try to regiment that which will not be regimented. Yet their activities are wholly good, so long as the food produced is of the right quantity for a limited population. But of course it is not. The same race of chemists who help farmers with artificial fertilisers, insecticides and weedkillers have put down yellow fever, plague, leprosy and the like. They have doubled the life span of a typical *H. sapiens* in a few centuries. Nature loathes an excess of any one species or of any one activity. If we begin to feel the pressures of over-population now, what might those pressures be in another half century? Our cup will 'run over', and nature, rather than us, will put it right. Our assessment of 'us' still includes, among those who rule, a feeling that we can work out our own destiny. We are about to discover not so much the error of such thought but the extent of the error.

Ultimates

There was one other conclusion from the world conference of insects. They admitted that there was one facet of life at which our species really did excel. It was our ability to kill our own kind – sometimes as many as a million a year. We are a race of creatures that seeks extremes: the ultimate in care for the sick, in love of our children. But we are more vicious than the tiger that kills without cause. Each machine we invent that uses principles that *we* consider new, but which have been used by living creatures for millions of years, is invariably put into service first as a deadlier weapon of war, later for better purposes. The insects cannot understand the cause of most of our wars, for they know neither religion nor politics.

The Guinness Book of Records[5] is evidence enough that wherever humans are found they will be seeking the biggest, the smallest, the fastest and the highest. We *are* different from all other life on earth. We could build paradise beyond all our dreams if we blended biology and technology with 'playing to the rules'. What we do instead is well illustrated by the lyrics from the musical show *Kismet* based on Borodin's music:

> Princes come, princes go,
> An hour of pomp and show, they know.
> Princes come, but over the sands of time they go.
>
> Wise men come, ever promising
> The riddle of life to know,
> Wise men come but over the sands
> The silent sands of time, they go.
>
> Lovers come, Lovers go,
> And all that there is to know
> Lovers know – only Lovers know.

'The tragedy of youth,' wrote G. B. Shaw, 'is that it's wasted on the young.' If some of us have known the answer to our problems when we were young, why did we join 'the mob' as we grew older?

6

Topology – the master discipline

What is 'matter' and what merely 'shape'?

In the pursuit of what we call 'truth', the question: 'What is "stuff"?' always stands at the end of the road. It stops pure scientists dead in their tracks. It confuses the holy people. In scientific circles, hypothesis follows hypothesis and concepts merge with each other in a way that suggests 'infinity' in whichever direction one looks. For when a concept of a 'smallest bit' of matter is formed, it is at once destroyed by the argument that if the mind can imagine a smallest solid bit, that same mind can imagine it being split in half, so it is no longer the smallest! What sticks in the gullet is the idea that everything solid must have a shape. Yet clearly not all things that affect our physical bodies are solid (the word 'solid' here includes liquids and gases, which is why 'stuff' is a better description of it). Heat, for example, is none of these and has no shape. But heat is just a form of energy is it not? Careful – energy is a concept, a thing of the mind, a relative thing. If two space-vehicles pass by each other, how shall a traveller in either of them know which of the two has the greater energy?

Radiant heat is only one kind of radiation, defined only by its frequency, the others being radio waves, visible light, ultra-violet, X-rays, gamma-rays and 'beyond . . .'. Radiation is transmitted as waves. 'Waves in what?' stands in our would-be path to truth.

'Waves of probability' comes the scientific answer, which satisfies very few of us. What makes a lot more sense is that radiation on the one hand and stuff on the other could well be one and the same thing, the difference depending on how we choose to look at them. To accept stuff as absolute, unconvertible stuff would be easy. It would push aside the question: 'Waves in what?', but those who choose to answer the latter question have a deal of explaining to do in connection with two bits of matter, each smaller than the nucleus of a simple atom, that disappeared in August 1945 and with the fact that as a result two Japanese cities were wiped out with devastation such as the world had never seen before.

Yet to accept the easy concept of stuff, the question still gnaws at us: 'What shape were the bits that disappeared?' If the answer comes, as it inevitably must, that there are 'lumps of matter that have no shape', we are forced to the conclusion that shape is a concept of the mind only. In this respect it is in good company, for its fellows are magnetic and electric flux, magnetic poles and electric charges, gravitational waves, and perhaps even mass itself. For every one of these items cannot be described *absolutely*. They can only be *related* to something else and ultimately to oneself.

And yet, even these concepts of the mind have a pattern – a symmetry. The profound nature of these things was wonderfully expressed by Professor E. C. Cullwick in a book on electromagnetism in the 1930s.[1]

> The known facts of electricity and magnetism form an exact
> and coherent body of knowledge of surprising beauty and
> symmetry, and the unravelling of this beauty by patient
> thought and experiment forms one of the most fascinating
> stories of all time. From small and disjointed beginnings, from
> lodestone and amber, this most incorporeal of nature's
> secrets gradually capitulated to the restless mind of man until
> at last the knowledge handed on by Oersted and Faraday,
> by Ampère and Maxwell bids fair to embrace the whole of
> the physical universe.

Modern authors of scientific and technical papers please note: we are losing the art of expression in our struggle for scientific flawlessness. By comparison with a lot of the modern textbooks on electromagnetism, Professor Cullwick's writing is sheer poetry. On the subject of teaching its fundamentals he had this to say.

> How confusing it must be for (the student) if he is not told at once what this fourth physical concept is. If he is introduced to electric charges, magnetic poles, electric and magnetic fields as though they were all equal in the possession of physical reality, little wonder that they become in his mind all equal in mystery. And if he is then told, in dimensional formulae, that none of these is to be linked with mass, length and time in the quaternion of primary concepts, but that he must use *either* the 'permittivity' *or* the 'permeability' of free space, the race is indeed for the swift, and the battle for the strong.

Euclid and after

Whether we like it or not, the principles of Euclidean geometry (whether or not his teaching is followed to the letter) have been handed on from generation to generation, not so much because they were *useful* as that they represented a *discipline* which, according to school teachers, dictators and sergeant-majors, is vital to the life of any human! Latin and Ancient Greek are still being handed on for much the same reason. Euclid allowed only ruler and compasses for all constructions, thereby laying his own particular foundation stones for the whole concept of shape – the straight line and the circle. Lesser people then wasted considerable fractions of their life spans trying to succeed either in trisecting an angle using ruler and compasses only, or in proving that it could not be done (the latter, incidentally, triumphed, but who cared?).

Drawing lines with a ruler is easy, and the definition of 'straight' as being the shortest distance between two points follows as entirely palatable, until one probes the depths of space, asks for a meaning for space and is rewarded with the answer that space is related to matter (almost like going back to the start in a game of snakes and ladders), and that one must either admit the shortest distance between two points in space depends on what masses you pass by on the way, a concept giving rise to a geodesic,* or that space is curved in the presence of matter, which is tantamount to saying that your straight Euclidean ruler is bent if it is very long!

Among the problems that face any civilisation fairly early in its life is the fact that the relationship between the circumference and diameter of a circle can never be expressed in decimal (or any other system of whole numbers of) digits so that the series terminates or repeats. The best we can do is to give it a symbol, π, and to be prepared to work it out to any required accuracy, thus

$\pi = 22/7$

$\pi = 3.142$

$\pi = 3.1416$

$\pi = 3.14159$ . . . and so on.

You can have a lot of fun with ruler and compasses, drawing apparently curved lines with a ruler, the curve being the 'envelope' of many straight lines. But they will avail you little in appreciating the shapes of nature. The derivatives of Euclidean geometry in the matter of *solid* objects include cubes and brick shapes, rectangular beams, pyramids, spheres, cylinders, tubes and right circular cones; the last four became a 'ball and chain' of enormous size for humans, once they had discovered the action of a wheel.

---

* A geodesic is best described (as are most things, including the Kingdom of Heaven) by the use of parables ('analogies' in scientific language). The shortest distance between a town in England and one in New Zealand is through a hole passing through the centre of the earth. As none of us is likely ever to encounter such a hole, the information is of little value. But the shortest distance by aeroplane is around a 'great circle', where only the curved surface of the earth acts as a restriction. Such a path is a 'geodesic'. But before aeroplanes were invented the shortest route was by sea via the Suez Canal. That route also was a geodesic. When the Suez Canal became full of scuttled ships in November 1956, the sea route around the Cape of Good Hope was the geodesic. A geodesic is the shortest distance between two points when obeying a number of constraints.

The lathe is preferred to the milling machine in terms of pure accountancy in the latter half of the twentieth century especially, and the number of objects with circular sections that we encounter in a day in a city is very large.

To the question: 'Are not tree trunks, branches and stems of grasses, flower petals, berries, birds' eggs and many other natural things of circular section?', the answer is quite clearly: 'No! – only superficially.' The one thing that nature has in common with Euclid is a preference for symmetry, but only bilaterally, and often then, only superficially. A neat summary of the underlying cause of bilateral symmetry in animals is provided by Martin Gardner in the *Scientific American*:[2]

> It is easy to understand why this symmetry evolved. On the
> earth surface gravity creates a marked difference between up
> and down, and locomotion creates a marked difference
> between front and back. But for any moving, upright creature
> the left and right sides of its surroundings – in the sea, on
> the land or in the air – are fundamentally the same. Because
> an animal needs to see, hear, smell and manipulate the world
> equally well on both sides, there is an obvious survival value
> in having nearly identical left and right sides.

This is fine so far as external appearance is concerned. What is interesting to note is that our major machines that allow us greater freedom of travel, the motor car, the ship and the aeroplane, also have this symmetry where only the need to 'manipulate the world equally on both sides' is relevant. My colleague Gilbert Walton again, this time on the subject of bilateral symmetry.[3]

> Both natural and artificial things show an enormous variety
> of possible shapes. Jewelry and snowflakes have their perfect
> symmetry, towns and continents have none. The symmetry
> of highly elaborated things, both natural and artificial, is
> sometimes strangely similar. Outwardly in an animal, as in a
> motor car or ship, there is bilateral symmetry. Inside, however,
> the organs or components are not all arranged symmetrically;
> the heart is on one side in the mammal, and the carburettor is
> on one side in a car . . . In the final detail the loss of symmetry

is complete. The molecules of haemoglobin in the blood, or of enzymes in the muscles, are apparently specially fabricated in a highly irregular but precise manner just as the details, for example, of the electrical system of a vehicle are completely irregular and show no symmetry. The comparison can be pursued further. Our right hand is a mirror image of our left hand, but a glucose molecular for instance in the tissue of our right hand is not a mirror image of a glucose molecule in the tissue of our left hand. Both are 'dextro-rotary' and 'laevo' glucose plays no part. This complexity is the same in artificial things. The right side mudguard of a car is the mirror image of the left side mudguard, but nuts and bolts in both mudguards have right handed threads.

One could liken the nervous system or the blood vessels in a mammal to the electric wiring of a vehicle and, in the case of large aircraft, the complexity is within an order of magnitude of that found in nature, and we did all this without any conscious effort to copy nature. That we did it in entirely different materials, metals as opposed to organic chemicals, makes the comparison all the more striking. But, of course, we went further in the pursuit of more than one degree of symmetry. Manufactured objects that stop at one degree of symmetry include garden tools and cutlery and other simple instruments that are more utilitarian than 'beautiful'. Vessels for liquids that are fundamentally cylindrical, but have handles and/or spouts, have only one degree of symmetry (cups, teapots, jugs, etc.). Two-dimensional objects including all pictures on flat sheets of paper or canvas can have more than one axis of symmetry. A rectangle has two, but is not symmetrical about its diagonals, whereas a square has four axes of symmetry that *include* the diagonals. Hexagons have six axes of symmetry, and so on, until we approach the 'perfection' of the circle with an infinite number of axes of symmetry.

Special properties of special shapes

What we must consider, however, is whether the obsession with the square, the hexagon or the circle is more than a mere seeking of perfection or of beauty for its own sake. The rectangle that contains maximum area for a given perimeter is a square. The perimeter shape that encloses the ultimate maximum area for a given perimeter is the circle. Here is an economy that must be involved in the natural growth of things as well as in our engineering economics.

When we leap from two- to three-dimensional shapes, topology explodes. Axes of symmetry can still exist, but in addition, there can be *planes* of symmetry and points of symmetry. A cylinder, for example, has a single plane of symmetry Z equidistant from its ends, which is at right-angles to the axis of symmetry AB (see Fig. 6.1a). But there are an infinite number of planes of symmetry such as PQRS, each of which *contains* the axis of symmetry. It is not difficult to see that these essentially different symmetries are the result of idealising the solid so that its geometry parallel to one particular plane is always a circle, and the *same* circle, whereas its geometry parallel to any plane at right-angles is always the same rectangle.

The cube is a very special kind of solid belonging to the group known as 'parallelepipeds', of which the most general is shown in Fig. 6.1b. The definition of this group is fairly clear from this figure. It contains all solids having six plane faces, each of which is a parallelogram. On the way to a cube, of course, one comes to the 'brick' shape in which all faces are rectangles. A cube has 13 axes of symmetry, 9 planes of symmetry but no point of symmetry. In solids there are several kinds of axes of symmetry. The cube shown in Fig. 6.1c can be rotated about an axis such as AB so that it presents to an external observer four views that are indistinguishable from one another each revolution. But rotation about an axis, such as CD, produces only two such views. AB is called a 'fourfold' axis of symmetry and a cube has three of them. CD is a 'twofold' axis and a cube has six of them. A corner-to-corner axis such as EF produces threefold symmetry and of these there are four. A cube is the parallelepiped that

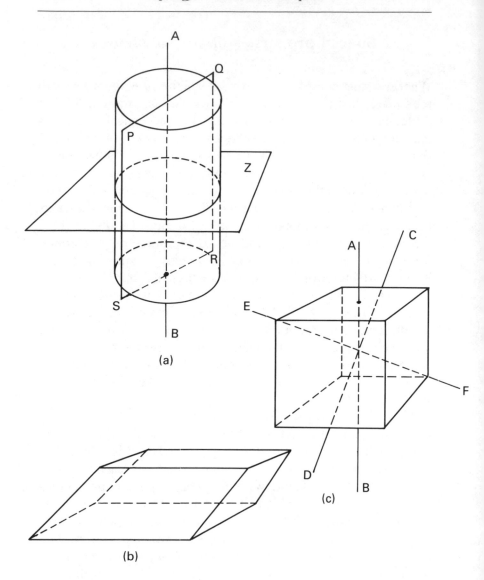

Fig. 6.1. Axes and planes of symmetry. (a) A cylinder. (b) A parallelepiped. (c) A cube.

contains maximum volume for a given surface area. But it is also clear that there are other solids bounded by planar surfaces that are more advantageous in this respect, e.g. the shapes that include those with special names, like the octahedron, dodecahedron and icosahedron. These are all *regular* solids in that each face is a regular plane figure having all its sides and all its angles equal. The advantage of regular figures is that if the solid itself consisted of no more than rigid fibres along the edges with thin films as the faces between the fibres, enclosing a fluid interior, then internal pressure would produce equal stresses in all the members, i.e. there would be no weak spots. What is interesting, however, is that something is built into our genes that gives us a preference for *regular* figures and in the ultimate one comes to the conclusion that things that *look* good are *functionally* good. This theme is much abused in the modern world, and 'streamlined' cars have become a status symbol, when in fact the improved performance due to the streamlining only begins at about the legal limits of speeds on the majority of roads. More than that, it can be a fake friend if it becomes a *rule*, for the ugly in nature often flourishes as successfully as the beautiful.

'Solids of revolution' are anything that can be seen to start life as a two-dimensional outline which is then rotated about a particular axis. Such shapes are almost a prerequisite to the products of engineering because of our insistence that the wheel is the cleverest thing of all. A second reason is concerned with simple economics. A century ago men had time to carve asymmetric objects and their womenfolk knitted and crocheted intricate patterns, but once organisation had set in, the accountant had no time for hand-made articles unless those making them would accept a handful of rice a day as sole payment for the labour. Rectangular parallelepipeds first, and later the products of the lathe, alone dominated the thinking of the controllers of technology. Figure 6.2 shows how simple geometrical shapes on a surface 'come alive' when made into solids of revolution. Notice, too, how choice of the axis of revolution is as important as the shape of the primitive outline itself. Figure 6.2b is of special interest for the torus is often seen as a fundamentally different, three-dimensional shape from that of a sphere, for circular paths can be drawn on the surface of the torus that are very different from each other. This cannot be done on a sphere. Yet both shapes have the same two-dimensional ancestor.

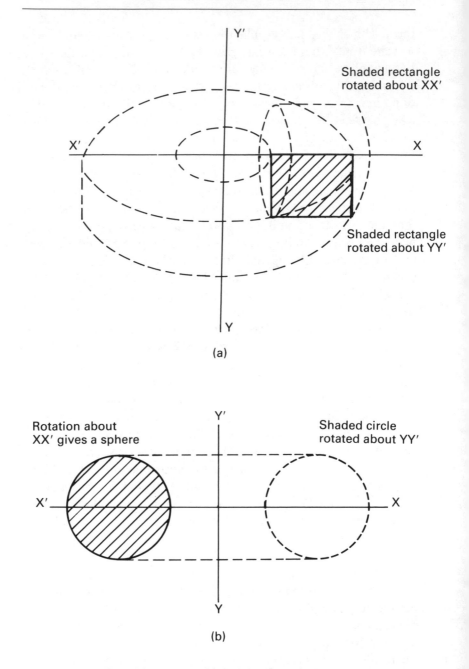

Fig. 6.2. (a) and (b) Solids of revolution.

Helices

Right-angles make up the basic structure of one form of topology, circularity/sphericity is another, but helices are in a way a combination and extension of each. Almost as far back as one can go archaeologically, one finds the helix as a symbol of mystery, of magic (black or white), of God, of fertility, and so on. Jill Purse, in her book *The Mystic Spiral*,[4] observes: 'Many formations in nature, although both constituted and caused by dissimilar phenomena, are not only similar to look at, but have identical mathematical description. This would suggest that together they form a higher overall order outside that limited by our concept of linear cause and effect.'

There are two-dimensional spirals of which the Archimedean spiral (Fig. 6.3) is an example. In three dimensions there are helices of fixed pitch and circularity – as in a coiled spring (Fig. 6.4a) – spherical helices, both re-entrant and non-re-entrant (Figs. 6.4b and c), and many others. Double spirals in a plane (Fig. 6.4d) were carved by Megalithic people. A helix around a torus (Fig. 6.4e) is the way in which Michael Faraday wound wire around an iron anchor ring in order to make the world's first transformer (1831). It is also the path each of us makes around the sun in the course of a year, although the plane of each of the 365¼ turns (insofar as one can speak of a 'turn' as if it were actually contained in a plane) is inclined to the circular axis of the torus in this case.

Tapered helices are common in natural things, especially in sea-shells. It is perhaps here, more than anywhere else, that we can see the helix as a natural *growth* shape. What can, of course, be a growth can similarly be a decay, whilst vortices in air (made visible by the addition of smoke) and in water are common in nature, the former in its extreme form is the whirling tornado. Both the water vortex and the tornado are capable of exerting enormous linear forces on objects in the vicinity of their axes.

Climbing plants should have shown humans the secret of spinning yarn long before it *was* discovered, ancient though the art is. The 'engineering' of a spun yarn is easily tested. Take a short

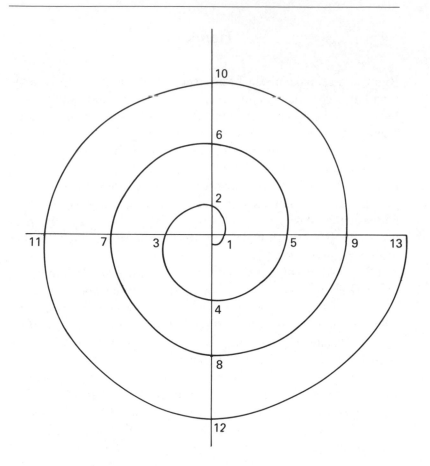

Fig. 6.3. An Archimedean spiral.

piece of sewing machine cotton and untwist it. If it consists of several 'ply', you will need first to untwist the strands, then take a single strand, discover the handedness of the twist in it and then untwist. As you do so, pull gently apart and discover that the material has lost **all** its tensile strength. The strength lay in the twist, *not* in the material. Surprising though it is, all the forces that resist longitudinal tension come from friction between fibre and fibre. But the secret really lies in the helix, for its nature is such that if the fibres slip at all, the result is to increase the pitch of the helix and to reduce its diameter. This forces the individual

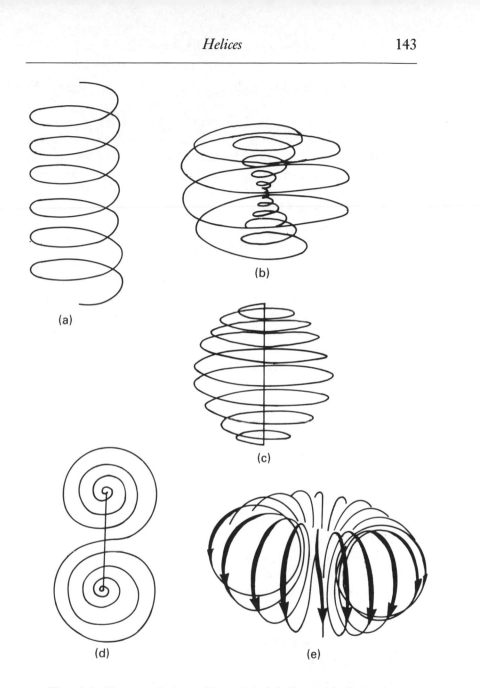

Fig. 6.4. Two- and three-dimensional helices. (a) Constant amplitude and pitch. (b) Re-entrant spiral. (c) Non-re-entrant spiral. (d) Double spirals in a plane. (e) A toroidal helix.

fibres against each other, radially, much more strongly, and the tensile strength is at first increased rather than decreased. Almost all ductile substances, including many metals, yield to initial tension without strengthening. The forces exerted on an electric wire due to the passage of current are such that they are radially greatest at the thinnest part, so as to make it thinner, and from the instant of yield the system escalates until the wire is ruptured and the two ends are blown apart by the explosive nature of the spark that follows the break. This is a description of the action of a fuse. It is also an example of 'positive feedback', of which more on p. 192.

Now that we can look at the surfaces of objects in the greatest detail, we can appreciate that the smoothness of a polished surface or of the petal of a flower is not smooth at all (see Fig. 4.17). Surface contact is, therefore, at best like the rubbing together of two ploughed fields, face to face! – small wonder that lubricants are so important in engineering. In the earliest of experiments with electricity, pieces of amber* were rubbed on animal fur to charge them so that they became capable of exerting small mechanical forces on other insulating materials. The belief that the energy which moves objects by this electrostatic process came from the frictional energy used in the actual rubbing has persisted into the twentieth century. Since, alas, this totally false information is still being handed out, a short digression here into the mechanism by which one can receive a quite painful electric shock on getting out of one's car is perhaps appropriate.

When any two different substances come into contact, the loosely bound electrons at the contact face have a habit of changing sides. What is more, each substance has its own inherent ability to release or to attract the electrons from the other substance. An analogy is the way in which two liquids, separated by a porous membrane, pass through it in different quantities in different directions and are able thus to create a pressure difference between the two sides (osmotic pressure). In the case of

---

* The Greek word for amber ('elektron') gave its name subsequently to the whole of electric (electronic) phenomena and therefore to the profession of electrical engineering.

electrons, they create an electrical pressure (voltage) that is given the name 'contact potential'. It occurs with metals and non-metals alike. The only thing about metals is that they are free to conduct the excess electrons, or the extra electrons needed, from or to the joint. Insulators retain them *in situ* and the voltage can be measured. Generally it is of the order of 0.5 to 2.0 volts. Now, a voltage is literally a *pressure*, i.e. a *force per unit area*. So if only a small area of contact is established, the voltage is the same as for a large area but the force is tiny because only a few excess electrons have accumulated. The reason for the rubbing of the amber is now clear. It is only a question of moving the two jagged surfaces about (jagged at micro level that is) to ensure that a large area of one comes in contact with a large area of the other.

In the electrical business, the ratio of the charge q transferred, to the voltage resulting from the transfer (whether by contact potential or by any other method, such as chemistry, light, magnetism, heat, etc.) is given the special name 'capacity'. (With the onset of Système Internationale (SI) units, Napoleon, Charles de Gaulle and all other 'rationalisers', the name has become 'capacitance', merely to rhyme with 'resistance', believe it or not!) An object with capacity used to be called a 'condenser' (due to the same orderliness it is now a 'capacitor'), but the reason for the idea of volume in a vessel holding a liquid is fairly obvious. One could say that if a beaker of uniform cross-sectional area A, and height h was filled by a quantity (volume) of liquid Q, it had a capacity C given by $C = Q/h$. In this context, of course, C is simply the cross-section A. With a condenser, Q is the quantity of *charge* poured in and h corresponds, in a sense, to the voltage, in that it is this quantity that will eventually set the maximum quantity that can be poured in. Thus, in electrical terms

$$C = \frac{\text{charge}}{\text{voltage}}$$

Now, in the context of electrostatics, a human may be regarded as a *conductor*, like a metal. (More correctly, an animal is an electrolyte – a collection largely of salts in solution.) A human sitting on a modern car seat (covered with highly insulating plastic) inside what is an almost complete metal shell (the car body) makes a very good condenser with a capacity of perhaps 10^{-12} Farad. (The unit of capacity is a very impractical choice. In micro-electronics, a million-millionths of a Farad is considered 'large'.) If now the human wriggles about on the car seat, the equivalent of rubbing the cat's fur on the amber, a considerable charge can be built up on the human. But at contact potential of 1.5 volts, who cares? Then the human happens to get out of the car *without holding on to the metal part of the car*. Their capacity could now be reduced by a factor of 10 000 or more, and since they have lost no charge, Q, the equation $Q = CV$ tells us at once that if C goes down by this amount, V must rise in proportion, and the 1.5 volts becomes 15 000 volts! When a finger is presented near the bodywork, a spark will jump when the distance between finger and car is such as to make the voltage per centimetre 30 000, in this case therefore at 0.5 cm (0.2 inch). I have known people generate sparks over 3 cm (1.2 inches) long by this technique. Avoidance of the shock is easy – just hold on to the metal as you get out!

You see, it was all a question of making large areas of one surface come into contact with large areas of another, whether it is for the collection of charge or for the accumulation of frictional forces. The tendril of a plant wrapped helically around a 'host' twig is self-tightening when a load is applied. (Try it with thin string around a pencil to see just how many turns are needed to obtain total grip.) Remember too, that as a plant grows its surface can be forced into molecular-sized grooves in the other surface so that often a single turn will suffice to bind them together.

Now, there are plants that entwine themselves around any slim object that they happen to touch. If the rotation is clockwise when viewed from above, there is more than a temptation to declare that the effect is merely the result of the tip of the plant following the sun (phototaxis). This is fine provided there is enough sun, which in a climate such as we experience in Britain is always doubtful. Still, one must be open-minded for in Von Frisch's famous bee-dance investigations, it was revealed that bees could determine the position of the sun in 10/10 cloud by the polarised light from the sky. So perhaps the plants can do this, too. That there is more than phototaxis at work, however, is evidenced by the fact that some species of plant go anti-clockwise whilst others do some of each! In the main, the mechanism is as simple and basic almost, one might say, as that by which a newborn baby grasps anything that touches the palm of its hand. It was just *built* that way.

Tendrils for climbing are obvious, even though we may never have applied any thought to the science of it. But there are other reasons for the formation of helices in plant life. Peter Stevens, in his masterly work *Patterns in Nature*,[5] uses the staggering theme throughout, that it is the nature of space that determines the shape of all things. Most of us dismiss the idea of space as a 'nothingness' – uniform in all directions – just a 'box' to put matter in. Stevens' opening page contains the beginnings of fantastic thoughts. The nature of space might not be as uniform as we think.

> In matters of visual form we sense that nature plays favourites.
> Among her darlings are spirals, meanders, branching
> patterns and 120° joints. Those patterns occur again and
> again. Nature acts like a theatrical producer who brings on the
> same players each night in different costumes for different
> roles. The spiral is the height of versatility, playing roles in
> the replication of the smallest virus and in the arrangement
> of matter in the largest galaxy.

Then, at once, a second idea, abhorrent to those who extend the theory of relativity to the thought that size itself is only relative.

> A look behind the footlights reveals that nature has no choice
> in the assignment of roles to players. Her productions are
> shoestring operations, encumbered by the constraints of
> three-dimensional space, the necessary relations among the
> sizes of things, and an eccentric sense of frugality. In the
> space at nature's command, five regular polyhedrons can be
> produced, but no more. Seven systems of crystals can
> be employed but never an eighth. Absolute size decrees that
> the lion will never fly nor the robin roar. Every part of every
> action must abide by the rules.

What a paragraph just 'for starters'! When I first read it I realised that it embraced all my own thoughts set out in the 'rules of planet earth', the 'rules of god' and the effects of scaling (Chapter 3). Stevens summarises his philosophy in two crisp sentences.

> Of all the constraints on nature, the most far-reaching are imposed by space. For space itself has a structure that influences the shape of every living thing.

In elaboration of this, he wrote:

> We did not recognise the special character of our space until the non-Euclidean geometers of the nineteenth century and Einstein in the twentieth century showed that there are other spaces, and that patterns and forms in these spaces differ from the ones we see in ours.
>
> Mendeleev, you see, did not say that space was *filled* with little particles, but that it *was* little particles. P. A. M. Dirac, John A. Wheeler, and other physicists have developed Mendeleev's idea and have likened space to a perhaps infinite array of tiny grains or a froth of bubbles. Perhaps, somehow the shifting of those grains or bubbles produces the fundamental particles that form the basis for all material structures.

Wheeler put it this way: 'There is nothing in the world except empty curved space. Matter, charge, electromagnetism, and other fields are only manifestations of the bending of space.'

These ideas are not as 'modern' as might at first be supposed. Leonardo da Vinci, Galileo, Newton, Michael Faraday, Osborne Reynolds and others were so far ahead of their time that we do well to go through their works with a fine-tooth comb to see whether or not we ourselves are sufficiently advanced to appreciate just what they *did* say. For example, Faraday wrote in 1845:

> A few years ago magnetism was to us an occult power, affecting only a few bodies, now it is found to influence all bodies and to possess the most intimate relations with electricity, heat, chemical action, light, crystallisation, and through it, with the forces concerned in cohesion; and we may, in the present state of things, well feel urged to

continue in our labours, encouraged by the hope of bringing it into a bond of union with gravity itself.

But now consider Faraday's conclusion about space.

> Space must be taken as the only *continuous part* of a body so constituted. Space will permeate all masses of matter in every direction like a net, except that in place of meshes it will form cells, isolating each atom from its neighbours, itself only being continuous.

Yet of atoms themselves, Tyndall had this to say of Faraday's Friday Evening Discourse at the Royal Institution on 19 January 1844:

> In this discourse he not only attempts the overthrow of Dalton's Theory of Atoms, but also the subversion of all ordinary scientific ideas regarding the nature and relations of Matter and Force . . . Like Boscovich he abolishes the atom, and puts a 'centre of force' in its place.

Such thoughts are as near to Stevens' brilliant exposition as perhaps anything that was written in the intervening period.

The movement of fluids

Theodore Schwenk's book *The Sensitive Chaos*[6] was based primarily on the movements of air and water in 'natural' situations. One special message runs throughout this book. Fluids are the link between the animate and the inanimate, a sentiment ('sentiment' it is, for unlike Moore's work it does not have the mathematical accuracy built into its many comparisons) expressed most beautifully in the Preface, which is written by Commandant Jacques Cousteau who has devoted his life to the underwater world.

> Gravity – I saw it in a flash – was the original sin, committed by the first living beings who left the sea. Redemption would come only when we returned to the ocean as already the sea mammals had done . . . All that life around us was really water,

modelled according to its own laws, vitalised by each fresh
venture, striving to rise into consciousness.

The author, in his foreword begins with accusations:

> Men gradually lost the knowledge and experience of the
> spiritual nature of water, until at last they came to treat it merely
> as a substance and a means of transmitting energy . . . The
> more man learned to know the physical nature of water and
> to use it technically, the more his knowledge of the soul and
> spirit of this element faded . . . A way of thinking that is
> directed solely to what is profitable cannot perceive the vital
> coherence of all things in nature.

Well, an engineer thinks just like that. I am an engineer. But I
try very hard to see the whole of my profession's activities in
nature, as witness the fact that I try to write this book which, at
best, can be only a glimpse of what lies ahead. Schwenk tells us
that:

> A prerequisite for an effective practical course of action is
> the rediscovery in a modern form of the forgotten *spiritual
> nature* of those elements whose nature it is to flow.

Many will argue that Schwenk has an obsession for fluid flow,
but there can surely be no doubts about this sentence: 'Through
watching water and air with unprejudiced eyes, our way of thinking becomes changed and more suited to the understanding of
what is alive.'

I myself have long considered a spinning wheel to be a 'live'
thing in that its behaviour when moved as a whole is very different
from the movement of the same wheel when it does not spin (a
'dead' mass). Schwenk at his best is doing no more than to suggest
that living things that came out of the water, were born in water
and/or lived in water have become shaped by the natural movements in their environments and that the resemblance between
a sophisticated sea-shell and a tapering vortex in water may be
much more than superficial. The photographs in Schwenk's book
are extremely persuasive, especially those comparing a ball and
socket joint on the one hand, and the bark of a tree on the other,
with fluid movements.

One further point about liquid vortices. Schwenk points out that a splinter of wood circling within a water vortex will always be oriented in the direction in which it first entered the vortex. In this it resembles a free gyroscope which will maintain its axis of spin in space to co-ordinates that can only be drawn with reference to the fixed stars that have not moved in the history of man. Schwenk's description of the wood in the water is remarkably similar, as he points out: 'This shows how a vortex is oriented – as though by invisible threads – with respect to the entire firmament of fixed stars.' Since the wood is not spinning, its direction-holding property must be entirely due to the motion of the fluid in which it is contained. What is strange is that if we trace out the path of each end of the splinter, we get a double helix. It can be shown that if a gyro consists of a wheel at the end of a long shaft, its 'natural' movement is such that the geodesics of the wheel centre and of the opposite end of the shaft also form a double helix. It thus appears that the wood in water comes very near to being a complete geometric inversion of a gyro spinning in a stationary 'fluid', i.e. the ether.

Basic natural shapes

In a classic work completed in 1917 and now available in a paperback, abridged and revised edition,[7] D'Arcy Wentworth Thompson wrote on the subjects of growth and form. Its title was simply *On Growth and Form*. His criticism of zoologists of half a century ago is happily less relevant to those of today.

> Even now the zoologist has scarce begun to dream of defining
> in mathematical language even the simplest organic forms.
> When he meets with a simple geometrical construction, for
> instance in the honeycomb, he would fain refer it to psychical
> instinct, or to skill and ingenuity, rather than to the operation
> of physical forces or mathematical laws; when he sees in
> snail or nautilus, or tiny foraminiferal or radiolarian shell a
> close approach to sphere or spiral, he is prone of old habit

to believe that after all it is something more than a spiral or
a sphere, and that in this 'something more' there lies what
neither mathematics nor physics can explain. In short he is
deeply reluctant to compare the living with the dead, or to
explain by geometry or by mechanics the things which have
their part in the mystery of life.

He continues, a paragraph later: 'Some lofty concepts, like space
and number, involve truths remote from the category of causation;
and here we must be content, as Aristotle says, if the mere facts
be known.' And then: 'Still, all the while, like warp and woof,
mechanism and teleology are interwoven together, and we must
not cleave to the one nor despise the other; for their union is
rooted in the very nature of totality.' – cosmology, no less, in the
1920s.

D'Arcy placed a lot of emphasis on the effects of surface tension, observing that: 'From the fact that we may extend a soap
film across any ring of wire, however fantastically the wire be
bent, we see that there is no end to the number of surfaces of
minimal area which may be constructed or imagined.' But he
pointed out that if such surfaces were limited to surfaces of revolution, there are only six (a fact first isolated by Plateau). These
six are the plane, the sphere, the cylinder, the catenoid, the unduloid and the nodiod. Each of these is generated by a curve AA'
being rotated about an axis XX' and each of the curves can be
generated by rolling one of the conic sections along that same
axis XX', like bowling a hoop, and marking out the path of the
focus.

Consider first when the conic section is a circle. Its focus is
its centre and this traces out the straight line AA'. AA' revolved
about XX' gives a cylinder (Fig. 6.5a). When an ellipse is rolled
its focus traces out a wavy line, which when rotated about XX'
produces the unduloid (Fig. 6.5b). A straight line in this context
is considered as the ultimate in ellipses whose focus may be
considered to be at the centre. This produces semicircles that
lead to a chain of spheres, like a string of beads. If the focus is
considered at any other point, the beads become alternately large
and small. Likewise, the parabola leads to the catenoid (Fig. 6.5c),

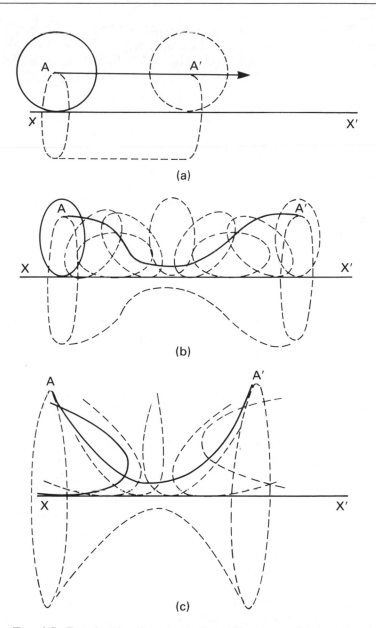

(a)

(b)

(c)

Fig. 6.5. Revolution of conic sections about axes. (a) A circle produces a cylinder. (b) An ellipse makes an unduloid. (c) A parabola makes a catenoid.

the generating curve being a 'catenary', which is the shape in which a heavy chain or flexible rope will hang from two points at the same horizontal level. The rolled hyperbola produces a shape other than a catenary, containing a loop, and its surface of resolution was named a 'nodoid' by the discoverer of the surfaces, Delauney in 1841.* D'Arcy goes on to describe how to produce these surfaces by using globules of the oily liquid orthotoluidene which does not mix with water but has precisely the same density as water when both are at 24 °C. The shapes are then formed from the red translucent oil globule using a blow-pipe. Today this system has been commercialised into decorative lamps for the home in which a light below the water vessel both heats the water and illuminates the ever-rising and falling 'natural' shapes of the oil.

In all of this, D'Arcy Thompson was making the point that these mathematically connected shapes are bound to appear in organisms that are freed from gravity by floating in a liquid of their own density: 'That spheres, cylinders and unduloids are of the commonest occurrence among the forms of small unicellular organisms or of individual cells in the simpler aggregates, and that in the processes of growth, reproduction and development transitions are frequent from one of these forms to another, is obvious to the naturalist.' I am certain that this is far from obvious among many naturalists whom I know personally!

D'Arcy's description of the oil blob experiments should be made compulsory reading for engineer and biologist alike and the experiments should be repeated in all school science courses for children of any age above six. The most unruly classes can be brought to a healthy state of concentration by the sheer 'power' of the experiment, what D'Arcy himself described as:

> . . . the 'materialisation' of mathematical law. Theory leads
> to certain equations which determine the position of points

---

* Liouville, J. (ed.) (1841). Sur la surface de revolution dont la courbure moyenne est constante. *Journal de Mathematique*, **6**, 309. D'Arcy also rec-ommends as reading, James Clerk Maxwell (a 'Patron Saint' of electrical engineers!) (1849). On the theory of rolling curves. *Transactions of the Royal Society of Edinburgh*, **16**, 519–40. It is almost an eerie experience to note how apparently interchangeable are the minds of the great men of learning.

in a system, and these points we may then plot as curves on a co-ordinate diagram; but a drop or a bubble may realise in an instant the whole result of our calculations, and materialise our whole apparatus of curves. Such a case is what Bacon calls a 'collective instance', bearing witness to the fact that one common law is obeyed by every point or particle of the system. Where the underlying equations are unknown to us, as happens in so many natural configurations, we may still rest assured that kindred mathematical laws are being automatically followed, and rigorously obeyed, and sometimes half-revealed.

I hope that readers will find my frequent use of D'Arcy Thompson both relevant and justified. His is the prime example of the incisive mind, probing to depths far beyond the capabilities of the instruments of his day and virtually anticipating the revelations of the electron microscope, of holography and the modern technologies of thin films, crystal growth, single molecular layer encapsulation and superfluids.

Plateau, whose work was as inspiring for Thompson as his is for me, solved the problem of fitting a minimal surface to the boundary of any given closed space by the use of soap films. The problem was originally formulated by the mathematician Lagrange. Mathematics will apparently long continue to be the 'glue' that binds the animate and the inanimate. Plateau's philosophy certainly re-appeared for me in my time at Manchester University in the 1950s when 'apprenticed' to the late Sir Frederic Williams who, after spending a semi-lifetime inventing many things electronic, including the development of the world's first commercial full-scale computer, turned his ingenuity to 'real' machines and invented 'the Logmotor', a brushless variable speed induction machine that was too complex for industry to build economically. The Logmotor lacked nothing in elegance and ingenuity and it is not necessary to understand its complexities in order to appreciate his deep-rooted understanding of physical laws when faced with the problem illustrated in Fig. 6.6. Machine A carried a large number of wire projections (X) that had to be connected to a second set of wires (Y) protruding from machine

B. But whereas Y was a system of uniformly spaced wires, X was logarithmically distributed and the problem was to define a starting point such that the total energy loss in the connectors was a minimum. More than this, the criterion was that the minimum loss was to be achieved with a fixed volume of connecting material. Clearly, an arrangement such as that in Fig. 6.6 could be bettered by shifting the system A to the right, but how far?

Knowing that the resistance of a connector was proportional to its length, l, and inversely proportional to its area, A, and that energy loss was proportional to resistance, the energy loss per connector was kl^2/V with k a constant and V the volume. F. C. Williams solved the problem by analogy, using the fact that the energy stored in an elastic string was $\lambda x^2/l$ where λ and the natural length l were constant and x was the extension. He modelled machines A and B as two strips of wood into which nails were driven at each point such as a, b, c, ... p, q, r, ... The strips were mounted in grooves so that each could only slide

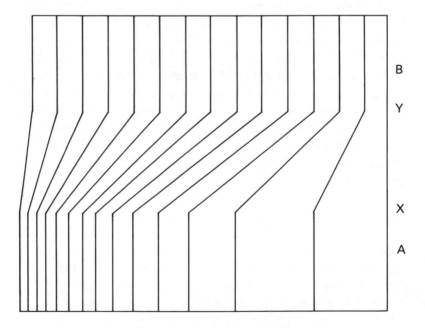

Fig. 6.6. Wiring diagram of a Logmotor.

along its own groove (horizontally in Fig. 6.7). Identical elastic bands were then made to join a to p, b to q, etc. and the whole system was released. Immediately it relaxed to the minimum energy situation that disclosed the pattern of connectors needed.

This example of the use of analogy reminds me to add to what is written on p. 20, a concise definition of the concept which I have never seen bettered. Originally credited to Quintilian,* it reads thus: 'The force of analogy is this, that it refers what is doubtful to something like it which is not in question.'

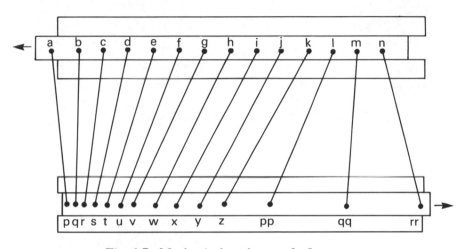

Fig. 6.7. Mechanical analogue of a Logmotor.

Basic natural shapes – bifurcation

In questions concerning randomness, nothing seems more haphazard than the ways in which trees branch. The general idea of increasing surface area by the use of leaves is an obvious development, but surely there need be no orderliness in the process? Strangely enough, the formation of branches has been the subject of more papers and book chapters than have other

* Quintilian was a first-century Latin teacher, writer and educationalist.

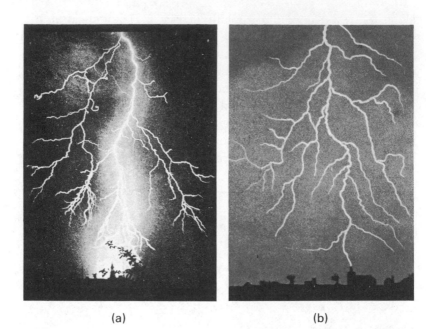

(a)   (b)

Fig. 6.8. Examples of branching.
(a) Is this forked lightning? (b) Or
is it this? (c) Or this perhaps?

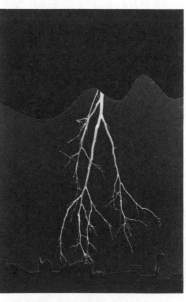

(c)

(d) This must surely be a plant. (e) And so must this. (f) And certainly this.

(d)   (e)

(f)

subjects involving natural order/disorder phenomena. Certainly the work of Richard Moore on mosaics is unique (see p. 209). The theory of entropy appears to deny any 'natural' movement towards order.

That regulated branching should be an organised process should have been as obvious to us as should many other engineering processes in nature. Invariably we fail to recognise the best ways of doing things until we ourselves stumble across them in our blind wanderings of technology that we then have the habit of declaring the work of genius. For tree branching is not unique to trees. Of the pictures that are Fig. 6.8a to f, only two are living creatures. One is a picture of forked lightning, one is the result of an electrical discharge through a dielectric medium and one is due to the seepage of manganese oxide through limestone. Of the two living things, one is a photographic negative of a tree in winter, upside down, possibly with the silhouettes of houses, etc. 'cheated' in below, the camera having been deliberately defocused slightly to give an impression of a lightning flash. The other is a portion of a small plant, after it has been pressed between the pages of a heavy book. The sixth is something quite different – could it be just a creation by an artist?

A careful inspection of Fig. 6.8a to c may result in the conclusion that none of these three is forked lightning. In the first, for example, there appears at the end of one limb, the beginnings of a helix. At the same side of the picture is an almost perfect figure 3, as it were, tacked-on. This could be the artist's attempt to confuse. What of b? It seems a little too angular to be lightning – and yet? Figure 6.8c has overlapping branches, almost certainly the result of seeing a three-dimensional setting as a two-dimensional picture, but do not be misled by this concept. Sometimes nature *chooses* two-dimensional forms.

In my professional capacity I teach my students of electrical engineering that electricity travels in straight lines when given the opportunity. Figure 6.9 shows a badly designed high-voltage laboratory where a sphere gap is being used to measure a high voltage. Sphere A is attached to the source of the voltage, S. Sphere B is earthed. The air between the spheres becomes conducting when the electric field rips electrons from the gas atoms,

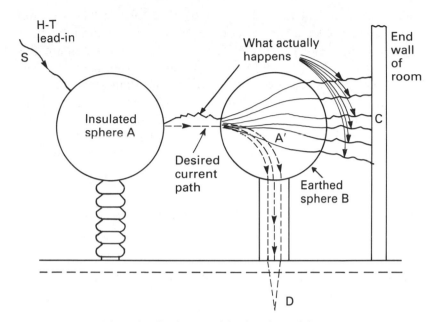

Fig. 6.9. A badly designed high voltage laboratory.

and the process rapidly escalates as showers of electrons bombard other gas atoms. But the designer of the high-voltage laboratory put the earthed sphere too close to the wall and when the explosive discharge occurred across AA′ it appeared to be carried onwards by a kind of inertia and emerged from the earthed sphere B and hit the wall at C. It objected to having to turn a right-angle at A′ in order to reach the ground at D, for to do so would be to traverse, as it were, a quarter of a turn of a coil of wire. Electricity objects to moving in circles.

So lightning ought to travel between clouds and the ground in straight lines, or at least in reasonably straight branches, so perhaps Fig. 6.8b is the lightning? Alas, no; Fig. 6.8a really *is* lightning, despite the helix and the figure 3 and the almost total absence of cross-overs, even though branching in three dimensions would appear to imply a high probability of cross-overs in depth. Figure 6.8c is the tree, upside down, and Fig. 6.8b is a map of the Amazon River!

Figure 6.8d is the other electric discharge, the man-made

lightning, but not through air. A block of plastic has been put under electric stress (top to bottom in the picture) and its surface punctured at one point (the base of the 'trunk'). Compare this with the plant (Fig. 6.8e) and then marvel at the nature of Fig. 6.8f, since this also is of entirely different origin.

These pictures do more than *suggest* that the bifurcated patterns are 'natural' shapes – as natural in their own way as the shape of a water droplet on a rose petal or of a snowflake which has a symmetry that surely belongs in a book of Euclidean geometry.

The theory of bifurcation

Bifurcation is a phenomenon that works 'both sides up'. Water droplets on a window pane will become too heavy to 'stick', run down the window, encounter an obstacle in the form of a dust particle or a tiny area of grease and be diverted laterally. Lateral movement means that neighbouring drops will coalesce, form heavier drops and proceed downward more rapidly. The process resembles that of the beginnings of a mighty river. Yet when the mainstream of the river nears sea-level, the huge volume of water spills out, bifurcating with the help of islands of silt that the river itself has carried down. Thus here, bifurcation occurs in the opposite direction – repeated splittings rather than repeated joints. A river delta has been formed. The effect is identical to that occurring as the result of growth.

Peter Stevens has dealt with the theory of bifurcation simply, yet, by the use of a good bibliography, completely. My treatment here differs from his only in detail. For example my Fig. 6.10 uses a packing of circles whereas his is one of hexagons. I have chosen to toss a coin to determine the moves left or right, whilst he used the digits of π, with the convention left for an odd digit, right for an even. The pattern of Fig. 6.10 is constructed as follows.

The centre of each circle of each row is to be joined to the centre of a circle in the row below it by a short, straight line (one only) so that a choice between left and right (heads or tails) has

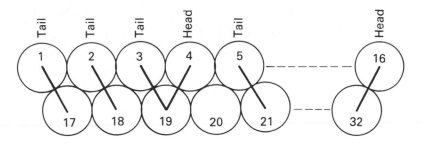

Fig. 6.10. A pattern generated by random choice.

to be made for each circle in turn (numbered 1, 2, 3 etc.) in the top row. When a row is complete, the next row is similarly treated in order. The result of my performing the head/tail sequence with a coin is faithfully reproduced as Fig. 6.11. Such patterns have counterparts in many facets of life, including sociology. To begin ascent from the lowest row carries no assurance of reaching the top, but to start from the top and to descend from *any* point is a journey that is bound to reach the bottom, without a possibility of making even a temporary unnecessary diversion.

Nature and engineering are both filled with ratchets and one-way valves. One may clutch a leaf of Pampas grass and draw the hand along it to its tip without harm. But a millimetre's movement in the opposite direction will produce slashes as severe as can be inflicted with a razor blade. Once the fluid has been pumped from the body of a newly emerged butterfly into the wing veins in order to stretch their 'canvas', the inlet pipe is sealed and the fluid can never return.

Referring again to Fig. 6.11, if we discover, by inspection, a complete network from top to bottom that is self-contained, it can be extracted and re-drawn in isolation. There are four such nets in the figure that do not involve spilling out of the picture sideways. Two of the more interesting of these are shown in Fig. 6.12a and 6.12b, where (a) corresponds to the top row starting points A to E, with a 'root' at X, and (b) to the row starting at F with a root at Y. Their appearance at this stage has only a superficial resemblance to that of trees, although even in so primitive a model there is more than a suggestion that (a) is a deciduous tree

Fig. 6.11. A tree/river pattern whose origin lies in random binary digits.

(a)

(b)

Fig. 6.12. (a) and (b) Two 'trees' extracted from Fig. 6.11.

whilst (b) is a species of conifer specially bred to be decorative.

An investigation of bifurcation such as the above, i.e. using random numbers, was recorded by Leopold and Langbein in 1962.[8] These were concerned with river tributary systems rather than with trees. An earlier paper by Horton on river topology,[9] subsequently modified by Strahler,[10] was perhaps the first attempt to put river patterns on a mathematical basis. It is worth noting that Horton was an engineer. The basic idea was to create an 'order' or 'hierarchy' in river tributaries whereby a 'first-order' stream comes from a source, such as a swamp, and has no other streams joining it. A 'second-order' stream is made up of the confluence of two first-order streams, subsequently added to by no more than first-order streams, and so on. It emerges that, statistically, first-order streams are about four times more numerous than second-order, second-order are about four times the number of third-order, and so on. Figure 6.13 shows Fig. 6.12a labelled according to this nomenclature. Remember that once a second-order stream is produced, subsequent additions of first-order streams do not produce third-order streams. Once a second-order has been produced, it stays as the same stream so far as the counting is concerned, until it is joined by another second-order. On this system, our 'deciduous' tree-type stream has 16 first-order branches, three second-order and one third-order. Considering what a small network sample this is, Horton's rule fits remarkably well, for he allows the ratio $4:1$ to imply 'anywhere between $3:1$ and $5:1$'. Notice also that the higher the order of stream, the longer it is, generally. Horton went on to show that the total length of all streams in a basin and the total drainage area of the basin could be estimated from the highest order number alone. Hence the length of a river's ultimate main channel (L) is related to the drainage area (A) by the formula $L = 1.4A^{2/3}$. In practice the formula was flexible to the extent that if written $L = KA^Y$, K varied between 1.0 and 2.5 and Y between 0.6 and 0.7 for a large number of rivers so analysed. Stevens points out that similar hierarchies of stream orders occur in things as diverse as lightning, cows' livers and economic market areas. What is most remarkable is that the same rules appear to work for rivers, human lungs and electric discharges, which are

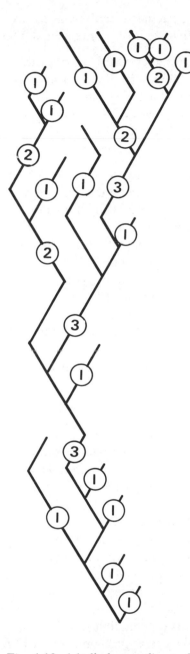

Fig. 6.13. Fig. 6.12a labelled according to Horton's rule.

essentially two-dimensional structures, as for tree branching and forked lightning which are three-dimensional. Stevens concludes:

> We begin to suspect, then, that Horton's statistical findings depend more on the general properties of space than on the mechanism of flow in actual streams – water is not the same as living tissue, nor is electricity the same as liver. The resemblances arise because in each of these systems the field of action, the spatial area, is the same.

It would be easy to accuse Stevens of 'leaning over' to make his philosophy work, as indeed it would be to treat Schwenk similarly and declare him *obsessed* with fluid flow, as was suggested earlier. Yet these gentlemen share very good company. Osborne Reynolds would have agreed with Schwenk point-by-point. Einstein's concept of matter was one of 'bent' space, as if the curvature of emptiness was sufficient to make it seem strewn with solid lumps. Even long before Einstein was born, Faraday had described space as if it contained 'lumps' (as quoted on p. 149).

Many decades were to pass before Rutherford established the electron, still more before Niels Bohr's model of an atom and Einstein's $E = mc^2$. Faraday would appear to be with us still as we argue that hydrogen nuclei cannot be solid. It has been said that there will always be a new sub-particle to be discovered so long as there are humans with enough money to look for it. Surely a new approach must come, and might that be not far removed from Faraday's concept of centres of force? It has a freshness that modern physics lacks. It is so full of what might even be commonsense. It sweeps away the useless search for that smallest bit of solid *stuff* (p. 131).

But to return to bifurcation. In creating tree-like and river-like patterns from random numbers, one of the sweeping assumptions was that all branches were at 60° to the vertical. In nature this is seldom the case. In 1926, Cecil D. Murray took Lagrange's principle of virtual work and applied it to flow problems, principally, in his case, to the flow of blood in arteries and their branches, but the principle is easily extendible to rivers and to trees.[11] Lightning too, is basically a flow problem, for as D'Arcy Thompson points out: 'The discovery of the circulation of the blood was implicit in, or followed quickly after, the recognition of the fact that the valves of the heart and veins are adapted to a one-way circulation.'

Murray's results might be said to be summarised in two figures (Figs. 6.14 and 6.15). The three different graphs refer to situations where, as in (a), a single branch P causes the main trunk T to deflect to T', or as in (b), a pair of branches P and Q leave

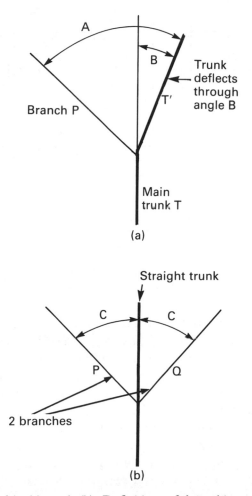

Fig. 6.14. (a) and (b) Definition of branching angles for Murray's analysis.

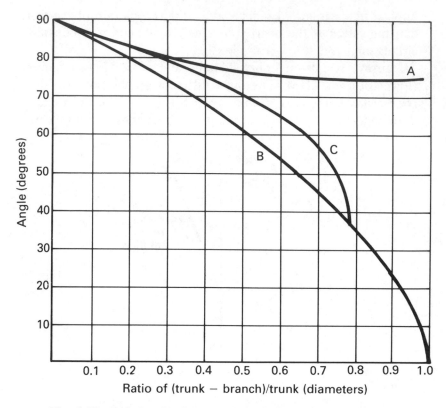

Fig. 6.15. Relationship between branch thickness and branch-ing angle (after Murray).

the direction of the trunk unaffected. It is interesting to note that Leonardo da Vinci had noted this fact: 'The branches of plants are found in two different positions: either opposite to each other or not opposite. If they are opposite to each other the centre stem is not bent; if they are not opposite the centre stem is bent.' Murray's graphs simply relate the ratio of the diameters of branch to trunk to the angles A, B and C. At a glance one sees that a single branch is always at an angle between 75° and 90° to the trunk. Stevens points out that one might, as a first guess, expect the total areas of branches and trunks to be constant at any level, implying a diameter ratio set by:

$$d_T{}^2 = d_T{}^2 + d_P{}^2 \quad \text{(using the notation of Fig. 6.14)}$$

but that Murray used the work function to take account of the distance travelled by the fluid against the resistance of the walls and therefore effectively used *volumes*, i.e.

$$d_T{}^3 = d_{T'}{}^3 + d_P{}^3$$

For me, the most incisive summary of the effect is made by D'Arcy Thompson:

> An approximate result, familiar to students of hydrodynamics is that the resistance is a minimum and the condition an optimum when the cross-section of the main stem is to the sum of the cross-sections of the branches as $1 : {}^3\sqrt{2}$ or $1 :$ 1.26. Accordingly, in the case of a blood vessel bifurcating into two equal branches the diameter of each should be to that of the main stem (approximately) as $\sqrt{(1.26/2)} : 1$ or (say) $8 : 10$.

He continues:

> Were blood a cheaper thing than it is we might expect all arteries to be uniformly larger than they are, for thereby the burden on the heart (the flow remaining equal) would be greatly reduced – thus if the blood vessels were doubled in diameter, and their volume thereby quadrupled, the work of the heart would be reduced to one sixteenth. On the other hand, were blood scarcer and still costlier fluid, narrower blood vessels would hold the available supply; but a larger and stronger heart would be needed to overcome the increased resistance.

If we now take our random-number-generated 'tree' (Fig. 6.12a) and apply Murray's curves to the angles of bifurcation and the branch thicknesses, as did Stevens for his primitive structure, we obtain Fig. 6.16, which is a great deal more like a real tree than is Fig. 6.12. But as Stevens points out: 'Real trees generate their parts in a regular and uncompromising manner and only end up looking random after disease and competitive struggle have taken their toll. Trees grow in a repetitive or modular manner. In any particular species, every bud is like every other bud.' The simplest module he suggests is a letter Y with equal

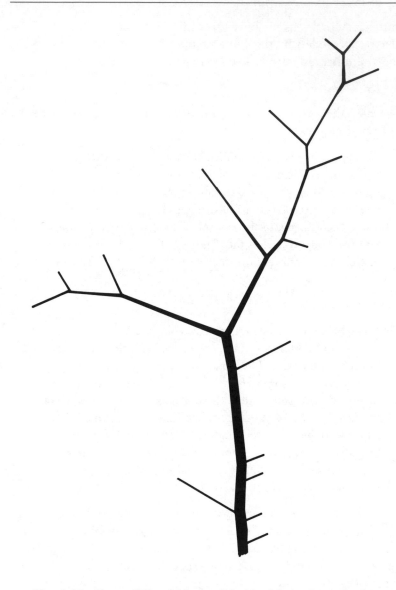

Fig. 6.16. Tree of Fig. 6.12a modified by Murray's method to include branch thicknesses and angles.

branches at 75°. A 'tree' constructed of four sizes of Y is shown in Fig. 6.17. But if a unit consists of an asymmetric module as in Fig. 6.18a, the resulting tree (b) appears conscious of a prevailing wind, or a sheltering wall. Stevens goes on to construct a modular tree from a simple Y with unequal arms, as in Fig. 6.19a, and produces Fig. 6.19b, and comments:

> All forks have equal angles. All forks have both a long and a short branch, with the long branch growing to the right from a long limb and to the left from a short limb. If you travel directly from the root to the extreme tip of any branch you will always encounter four forks. In short no arbitrary growth

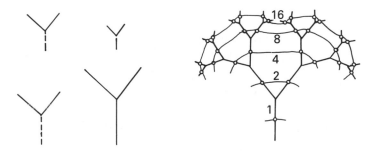

Fig. 6.17. A tree composed of simple Y modules (after Stevens).

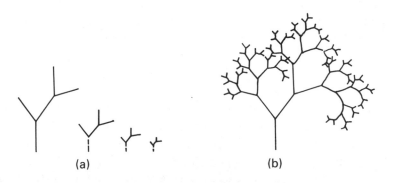

Fig. 6.18. A Stevens tree (b) composed of asymmetric modules (a).

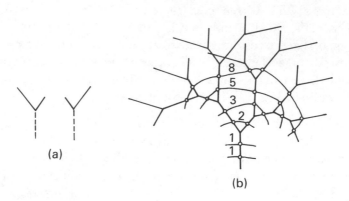

Fig. 6.19. A Stevens tree (b) composed of simple Y modules
with unequal arms (a).

occurs, and although simple rules are rigidly followed, note
how pleasing and graceful is the overall shape. The tree grows
from the regular combination of a module and its mirror image,
but appears neither so regular that it is dull nor so irregular
that it is completely amorphous. It falls in the narrow range
between order and diversity that we find beautiful.

Utopia, it appears, lies between perfect order and perfect chaos.
   The thin lines on Fig. 6.17 join the centres of each branch so
as to indicate 'growth contours'. The number of branches connected by each successive dotted line follows the simple binary
sequence 1, 2, 4, 8, 16, . . . , but if the same is done to Stevens'
unequal-branch tree (Fig. 6.19), the number sequence is 1, 1, 2,
3, 5, 8, . . . – our old friend the Fibonacci series (see pp. 204–
208)! Wherever there appears to be disorder, mathematics is never
far away.
   Stevens uses work by Ulam[12] to illustrate what is virtually the
opposite result to an order coming out of a disorder. What could
be more orderly than adding to an equilateral triangle three more
similar triangles, peak to peak, as shown in Fig. 6.20a? In Fig.
6.20b the process is repeated for a second 'generation' of triangles
mounted on every available corner. This is a pattern of perfect
order surely? But in Fig. 6.20c the process has reached the

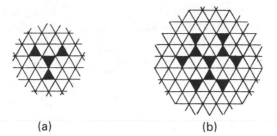

(a)　　　　　　(b)

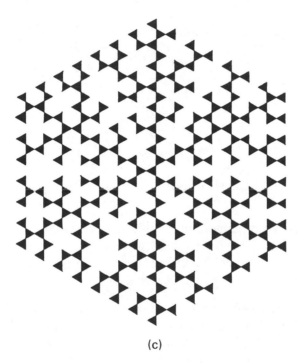

(c)

Fig. 6.20. (a) to (c) Ulam's expanding triangle patterns.

fifteenth generation and there is at least a doubt in the mind of the viewer as to just how many axes of symmetry this figure has. Stevens highlighted the system by replacing it with a network of lines joining the centres of all neighbouring triangles and produced the net shown in Fig. 6.21. It has six arms, each of which is a mirror image of the arm on each side of it, and at this point

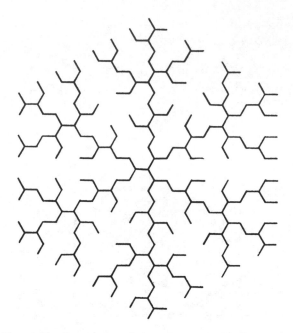

Fig. 6.21. Stevens' representation of a Ulam pattern.

the irregularity is no more than odds and evens. But Stevens goes on to develop the system for 28 generations and draws us just one arm of the set, as shown in Fig. 6.22. His commentary on it is better than anything I could write.

> It contains a definite central trunk, but its orderliness and the simplicity of the rule that gives it birth are not at all obvious.
>
> Like rivers and the patterns generated by random numbers, the tree has bifurcation ratios between 3 and 5, and its highest order branches are longer than the others. We also see the telltales of competitive struggles that we missed in the modular trees – the branches that do not bear fruit, that end before they have hardly begun. Statistically then, and visually too, the tree generated by a strict rule is similar to those generated by random numbers, and we observe that profoundly different models can portray the same phenomena equally well. Randomness can appear regular and regularity random.

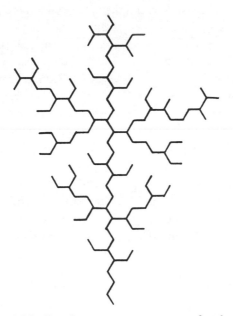

Fig. 6.22. Randomness appears out of order.

Where else can we find sentences to compare with this last? The work of Richard Moore comes to mind most readily from what was discussed on pp. 209–211, of course. But let us leap across the boundaries of scientific discipline and read the words of Sir Lawrence Bragg in some unpublished notes of which I was fortunate enough to be given a copy in 1966. He is discussing the beginnings of X-ray crystallography for which he and his father were jointly awarded the Nobel Prize. He is concerned with such profound concepts as Heisenberg's uncertainty principle, Bohr's atom, Planck's constant and the flow of time:

> The tidy structure of classical physics has undergone a great
> change and reappraisal during the last fifty years. It is
> generally considered that the first crack appeared in its
> armour when J. J. Thomson discovered the electron. 'J. J.
> Thomson opened the door to the new physics but never went
> through himself.' An observation which at first sight seems
> quite simple, reveals the need for a revolution in thought

when it is analysed. When light falls upon a metal, electrons are ejected, an effect known as the 'Photoelectric Effect'. This is natural; it would be anticipated that strong electric fields in the light waves would, as it were, jerk the electrons out of the metal. *But* whether the electrons are pulled out or not does not depend on the intensity of the light, but on its frequency. The feeblest light, of a frequency greater than a certain frequency characteristic of the metal, pulls out some electrons. The strongest light of lower than the critical frequency has no effect. If E is the energy required to extract an electron from the metal surface, the critical relation is $E = h\nu$ where ν is the frequency of the light and h is the famous Planck's constant. Conversely, if an energy change E occurs in an atom, which as a result emits light, the frequency of the light is given by $h\nu = E$.

This is a very extraordinary relationship. Let us take an analogy. Picture a lake with numbers of twigs floating on the surface. A twig is dropped on the lake from a height of one foot and ripples spread out over the surface. All the other twigs remain quiet, except one which may be a yard away or a mile away, and that one jumps exactly one foot into the air when the ripples reach it.

Energy is handed backwards and forwards from light to matter and vice versa in bundles or 'quanta' $h\nu$.

In fact light is behaving like particles, not like waves, as if it consisted of bullets which delivered up their energy $h\nu$ when they hit a target.

The story was rounded off by the discovery of electron diffraction by G. P. Thomson, and Davisson and Germer. Electrons scattered by matter exhibit interference patterns of a form which would be accounted for by their being waves. *Particles behave like waves, and waves like particles.*

The use of italics is mine. It is to emphasise just how similar is this sentence to Stevens' powerful observation. In science, it has been a repetitive story throughout its history that just when the ground seems most firm, the time for the earthquake to split the world is imminent.

7
Growth and decay

Change and decay in all around I see.
O', Thou who changest not, abide with me.
(H. F. Lyte 1793–1847)

The declining British butterfly

The inter-relationship between species of creature, whether they be plants, animals or insects, is too complex for any of us to attempt an understanding. But by examining more and more natural things we may at least get a 'feel' for the system and appreciate some of its marvels. Let us take a simple case of something that, superficially, resembles decay. In the 1930s a farmer would plough his fields to within six feet of the hedges that surrounded them. The border would be left to its own devices and grew lots and lots of weeds, in particular nettles. Now the nettle is the foodplant for a number of our commoner brightly coloured Nymphalid butterflies, the Small Tortoiseshell, the Peacock, the Comma and the Red Admiral. Came World War II, 1941 and the U-boat blockade, and Churchill's message: 'Dig for Victory' was rapidly implemented. A field that measured 200 yards by 100 yards plus a six-foot border had previously had a cultivated area of 20 000 square yards. Now the farmer ploughed it right up to the hedges and his potential yield increased by the factor $204 \times 104/200 \times 100 = 1.0608$, and 6% of all the crops in Britain would fill an amazing number of ships. Success in economics is made of just such changes. A sweet factory that wastes only $\frac{1}{2}$% of all its raw foodstuffs struggles to reduce it to $\frac{1}{4}$%, for then it makes an extra $\frac{1}{4}$% profits on its annual intake figure, and *that* can be millions of pounds.

But the butterflies of 1941 were cut down in numbers by much greater factors. The number of nettles in Britain was cut perhaps by a factor of 10. After the war, no farmer was willing to restore the 'wasted' borders; what good economist would do anything comparable? But you cannot have 6% more food and the same number of coloured butterflies. 'Civilisation' had taken another step – forward? In the years after the war, conservationism was effectively born and it spread like a fire. The conservationists deplored at length the 'disappearance of the British butterfly' – blamed practically everyone except the accountants, and got a lot of newspaper publicity. MBEs and the like were handed out for saving a single species of British butterfly that the majority of people had never seen anyway and the like of which is to be found in dozens of other species in other countries.

But when a butterfly population is divided by 10, so are the numbers of creatures that depend on those species for *their* livelihood. In particular, parasitic flies that lay their eggs under the skins of caterpillars, so that the larvae that hatch therefrom can eat the caterpillars alive, had a thin time too, in 1941. Their numbers also, when stability was restored, were divided by 10.

Now we are in a position to study what happens in a situation in which decay is inevitable.*

The laws of decay

Let us start with 512 units of whatever commodity is in process of decay and say that in every year one-half of all such units are destroyed. At the end of the first year only 256 remain. In the second year, one-half of these are eliminated, so only 128 remain.

---

* One feels that if there had been conservationists in the age of the dinosaur, they would have struggled to preserve *Tyrannosaurus rex* and would have failed miserably.

After three years, only 64 are left, and so on until, after nine years, only one is left. Figure 7.1 shows how this step-by-step decay takes place. It is at once recognisable as the number of teams left in the F.A. Cup after each round,* except that for the end of year 3 read 'end of round 2' etc.

But suppose we had subdivided the steps so as to assume that if one-half disappeared in one year, one-quarter would have disappeared in a half-year, and so on, the diagram would have been different (shown dotted in Fig. 7.1). Suppose we continued refining the technique until the steps were so tiny that we could not detect them with the naked eye and we would see only a smooth curve. What shape might that be?

Before attempting to answer this question, there are two others which must surely take priority. What justification have we for

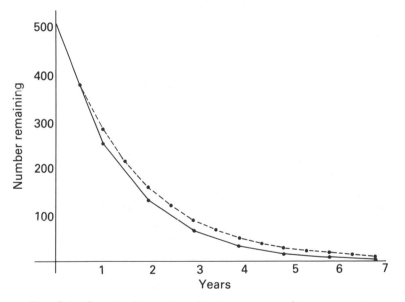

Fig. 7.1. Graphs illustrating the importance of time intervals in a decay process.

* This does not apply for the first two rounds where some of the less powerful teams and amateurs are thinned out to leave a net 64 at the end of round 2, no matter how many teams entered (provided the total number lies between 128 and 256).

assuming that decay in living things takes place according to such a rule? What happens when one reaches the last unit of a species? The latter is the easier to answer, so we will consider it first. The last dinosaur, or dodo, or great auk just dies. But in species such as one particular ant there are many more than 512 with which to begin and a million million million would not be an over-estimate of the present population. Even if they were of a decaying species (which many are not), the rate would not be 50% per annum, and long before they reached low numbers, the 'rules' would have changed, as we shall see. But in man-made things, such as in the rusting of iron left in the open, the rules are much less likely to change and the decay curve we seek theoretically becomes much more realistic.

Justification for continued decay at an ever-decreasing rate may be summarised by saying that decay can only attack *what is left*. Bullets cannot hurt the dead, and the dragging of Hector's body around the walls of Troy did no harm at all to Hector! Since 'what is left' is ever-decreasing, so must the overall *rate* of decay in terms of the number with which we began. Thus, in the example chosen, 50% (256) are eliminated in year 1. In year 2, 50% of what is left (128) are destroyed but these represent only 25% of the original 512. So the *absolute* fractions that disappear in successive years are $\frac{1}{2}$, $\frac{1}{4}$, $\frac{1}{8}$, $\frac{1}{16}$ ... etc. However, the next fact appears to go against our 'common sense': since there will always be 'some left' (if the population is infinitely sub-divisible), nothing, apparently can ever be eliminated. That this does not, of course, apply in cases such as great auks and football teams is due only to their not being infinitely sub-divisible.

'Natural' decay

To proceed with finding the 'natural' decay rate, Fig. 7.1 has indicated that the shape of the ultimate curve is largely set by the intervals that we set for changes in an otherwise constant rate of decay. For the full lines the fall rate was constant for a year at a time, changing as each year passed from a half, to a half of a half, $\frac{1}{2} \times \frac{1}{2} \times \frac{1}{2}$, $\frac{1}{2} \times \frac{1}{2} \times \frac{1}{2} \times \frac{1}{2}$, and so on. In the case of the dotted graph, the overall rate was a quarter for the first half-year, but the *next* half-year a quarter of what was left was not a half of a half but a quarter of three-quarters, and so on, so that the amount remaining was given by the series $\frac{3}{4}$, $\frac{9}{16}$, $\frac{27}{64}$, $\frac{81}{256}$ and so on, as opposed to $\frac{1}{2}$, $\frac{1}{4}$, for the year by year curve. Consequently, after a given time, in this case 2 years, more remains the smaller the interval we take.

Let us stop at this point in the argument and answer the question: 'Why is this numerical approach relevant to the extinction of a species?' Let us also take a highly simplified case to illustrate the answer; for example let us assume that a particular species of butterfly is being reduced by a collector. Each year it becomes more and more difficult to find them. The area of ground over which they are known to live stays the same. The 'density' of butterflies per square mile reduces all the time and hence so does the collector's success rate. It is not many years before it becomes necessary to spend a whole summer looking just for one! The same is true for parasitic flies and the butterfly's other enemies. So the natural decay curve really does apply. Now let us discover its mathematical nature.

By taking constant decay rates for a year at a time we saw above that the amount remaining after 2 years was $(\frac{1}{2} \times \frac{1}{2}) = \frac{1}{4}$, whereas to re-adjust the decay rate every half-year gave the amount left after 2 years as $\frac{3}{4} \times \frac{3}{4} \times \frac{3}{4} \times \frac{3}{4} = \frac{81}{256} = \frac{1}{3.16}$. We could obviously go on doing this; for example, taking every quarter-year would yield $(\frac{7}{8})^8 = \frac{1}{2.91}$, every month would be $(\frac{11}{12})^{12} = \frac{1}{2.84}$, until we reached the limit as the subdivision becomes infinitely large. This gives the amount left as $\frac{1}{2.718}$ and the number 2.718 is as important a number as π in any description of the human experience we call 'shape'. Like π, it is not calculable exactly in decimal digits. It can be shown to be the sum of the series of numbers:

$$1 + \frac{1}{1} + \frac{1}{1 \times 2} + \frac{1}{1 \times 2 \times 3} + \frac{1}{1 \times 2 \times 3 \times 4} + \ldots$$

as far as your patience allows you to go. To 10 decimal places $e = 2.7182818285$. It is given the letter 'e' (which stands for 'exponential') and it represents the natural decay curve (Fig. 7.2a) for all situations in which the rate of decrease of anything is a fixed proportion of the number remaining.

It is said that were civilisation, as we know it, to be wiped out, any new civilisation would 'rapidly' (say in a thousand years!) discover e, π, the square root of (-1) (which you know is impossible to evaluate) and another number called the golden ratio Φ, which we are to discuss later in this chapter.

Growth

Growth is a parallel process to that of decay. A good way of looking at growth is the way in which invested money accumulates interest. In 'simple interest' calculations it is only necessary to say that if £1000 is invested for 10 years at 10% per annum, the interest it accumulates is $(1000/100) \times 10 \times 10 = £1000$. But suppose an investor only deposits the £1000 for one year, then draws it out and re-invests it together with the interest that has already been paid. There is now £1000 plus £100 interest, i.e. £1100, to invest for the second year. In the second year this attracts $£(1100/100) \times 10 = £110$ interest. For the third year there is £1210 to invest, and so on, and at the end of 10 years, the total has risen to £2593.74 instead of the £2000 on 'simple interest'.

The investor is entitled to ask: 'Would I have got more if I had withdrawn the money every half-year instead of every year?', and of course the answer is 'Yes'. This looks like a bonanza. Suppose re-investment is made every week, every hour, every second, every micro-second – the final amount goes on increasing! But not, of course, indefinitely. What we are doing is to allow every £1 to increase by a fraction of itself (the fraction being the interest rate multiplied by the time for which it is invested) every period for which it is invested. We are therefore asking: 'What is the maximum value of $(1 + 1/x)^x$ as x gets bigger and bigger, and the answer is 2.718 or e, so our investor can never get more than £2718 by re-investing infinitely rapidly.

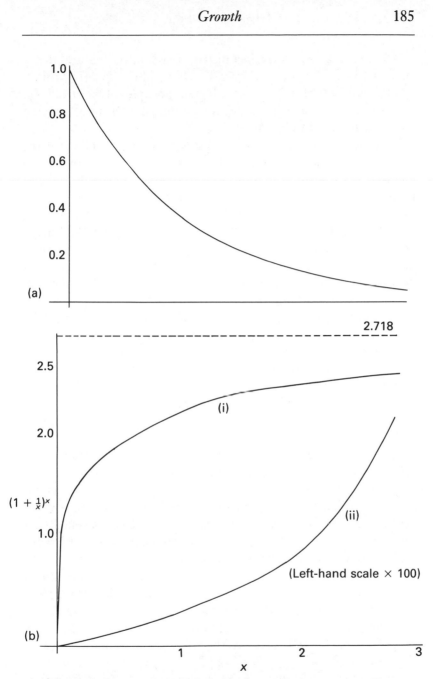

Fig. 7.2. (a) and (b) Natural (exponential) decay curve. For explanation, see the text.

The fact that this curve is a mirror image of our natural decay curve in Fig. 7.2a (shown as curve (i) in Fig. 7.2b is not an indication that this is a natural growth curve. It is simply a plot of the advantage of re-investment at ever-decreasing intervals. 'Natural' growth uses e in a quite different way.

Imagine a new species of creature introduced to planet earth from another galaxy. It breeds like a small mammal but has no enemies because it is inedible to all creatures on earth and not particularly attractive to look at. But it reproduces at one generation a year and has 10 'children' at a time, so the single pair that arrives have 10 offspring in the first year, 100 'grandchildren' after 2 years and in 10 years there are well over 10 thousand million (including remnants of the earlier generations not yet dead). This is unrestrained growth, where the rate of increase at any time is proportional to the number then present, so it is a curve of the same shape as the natural decay curve in Fig. 7.2a but as it were 'upended' (turned through 90°) as shown in curve (ii) of Fig. 7.2b. If there are as many people alive at any one time as have died in the whole of time previously, then the population is increasing at the natural 'e' rate. The fact that we are now exceeding this not only makes a nonsense of reincarnation but should warn us of the major problem that faces our grandchildren.

Unrestricted (explosive) growth of this kind cannot long persist, for fairly obvious reasons, but sometimes it can go unchecked for longer than is 'good for us'. An example of the latter is the atomic bomb, of which a beautiful analogue model was built many years ago in order to illustrate the explosive process. It consisted of a rectangular box of transparent material, on the floor of which were set hundreds of mouse-traps, side by side in rows and columns. A single steel ball was dropped through a hole in the roof on to one of the traps. As the trap went off it leapt in the air, landing on or disturbing several other traps and within two or three seconds the whole box was filled with flying mouse-traps! A simple and amusing display was the analogue of something deadly and terrifying.

Sir James Jeans' allocation of a profession to God (p. 88) is not surprising, for the laws of growth and decay are linked to numbers if they are to be regarded as linked to anything at all. But we must never lose sight of the fact that the human mind created numbers just as it created God, and growth and decay proceed unaware of either.

Saturation

One of the factors that has inhibited cross-fertilisation between the biological sciences and engineering technology is undoubtedly the different 'languages' that have been built up by the different groups of people at the frontiers of knowledge of their subjects. The biologist was unaware of what the electrical engineer called 'frequency response' for decades after the term had taken its place in technical dictionaries, mostly because it was used in the context of radio waves. The fact that bees can 'see' in a part of the ultra-violet was not appreciated as a description of the same phenomenon. The engineer for his part does not usually think of reading a book on ant behaviour when faced with a new problem in mechanical handling.

A phenomenon like saturation is generally regarded as 'belonging' to physics, because it is taught at school as part of 'electricity and magnetism'. The elementary physics book 'explanation' of magnetic saturation will suffice as a start, even though such a theory, first attributed to Weber in the last century, was superseded by a much more satisfying theory of domains (see pp. 79 –81). In Weber's simple theory, an unmagnetised piece of iron was said to consist of millions of tiny bar magnets lying higgledy-piggledy throughout the bar, as shown in Fig. 7.3. When an external magnetic field is applied it begins pulling the magnets into line, and the initial success rate is high. The magnetising field, H, is seen as the *cause* and the resulting magnetisation of the bar, B, as the *effect*, and the relationship between the two is usually plotted as shown in Fig. 7.4. From O to P one catches the merest glimpse of the explosive exponential rise of Fig. 7.2b(ii) as the magnets help each other to line up, and the more of them that are pulled into line, the more likely they are to pull those

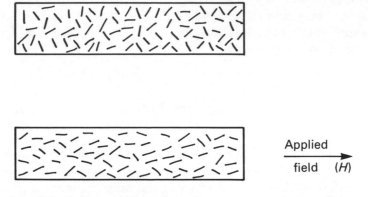

Fig. 7.3. Representation of a magnet as an aggregate of tiny magnets.

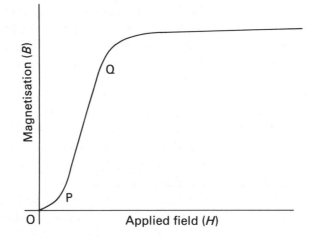

Fig. 7.4. The relationship between flux density, B, and magnetising field, H, for a sample of steel.

that are not. But the system becomes limited when it is mostly the external field doing the job and a linear relationship between B and H exists from P to Q, which is the usual phenomenon of proportionality, i.e. the bigger the cause, the bigger the effect. But you cannot have 'more of the cat than its skin', to quote an old-fashioned saying with strong biological connections (!), and you cannot pull more than *all* of the magnets into line. The growth curve OPQ begins to level out beyond Q and the iron is said to be 'saturated'. The word itself has its origins in chemistry, where you can only persuade a given amount of salt to dissolve in a given amount of water, and soon became a part of everyday language in that a person caught in a rainstorm without water-proof clothing could get so wet that their absorbent clothing could hold no more water. They were, in a sense, more than 'soaked to the skin'. Their clothing was *saturated*.

These examples will suffice to bring the gap between the common language in which 'you can't have more of anything than there is' and the scientific shorthand we use when we say something is 'saturated'. So where can we find an example of saturation in nature? An obvious one is a plague of locusts. Even before the advent of insecticides and of aircraft for spraying them *en masse*, a swarm of locusts only produced a *local* disaster. They never 'took over' the world. There was a limited amount of green food in any one area of land, a limited distance an adult locust could fly and a limited lifespan to each locust, and there had to come a time when the locust population was increasing faster than the green food could re-grow. This is 'saturation' so far as locust life is concerned.

But this is as far as the magnetic type of saturation phenomenon goes. Unlike the bar magnet, living processes such as the build-up of a locust swarm cannot stay at top level, for the food is consumable, and the next part of the locust story belongs to another phenomenon entirely, which is described in the next paragraph.

A description of another form of the saturation phenomenon is more readily applicable in some biological situations, such as the locusts, for it has a more complex and less obvious mechanism, and biological phenomena are not often simple. The thought was first put into my head by my (then) 13-year-old son who had seen an item on TV news that '1500 cars an hour were passing over the Avon Bridge' during a summer holiday traffic jam. I remarked casually: 'But *very* slowly.' 'Yes,' said young Dennis, 'but if there had not been so many people going on holiday there might have been 2000 cars an hour going over the bridge.' He *had* to be right – the more people trying to get over the river, the longer it would take any one of them to do it. But this only happens when the numbers are large. If only a few cars a minute wanted to cross, the time taken by each would be limited by the top speed allowed (either by *legal restriction*, or by *safety* consider- ations or by car *engine performance* in the last resort). Now, all three of the limits in italics represent saturation phenomena of the bar magnet kind and would give rise to curves of the form shown in Fig. 7.5 if one plotted 'cars per hour' in place of *B* and 'number wanting to cross' in place of *H*. But once the spacing between cars becomes important, human reaction time, rather than any of the three limits listed above, sets the maximum speed, and the slower the cars move, the closer they are packed and the more important reaction time becomes. The curve of Fig. 7.5 gives way to that of Fig. 7.6 and a single event, such as the failure of one car engine somewhere along the line, seizes the whole lot and we reach point X on Fig. 7.6.

This is a different kind of saturation in which the intensity of the result limits the power of the cause, i.e. changes the relation- ship between cause and effect. One encounters it in physics in

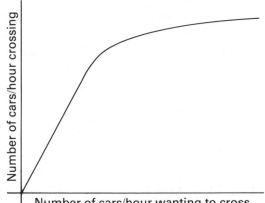

Fig. 7.5. An unusual saturation phenomenon relating to road traffic.

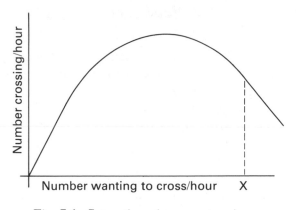

Fig. 7.6. Saturation gives way to seizure.

such things as superconductors* in which this special material can be so heavily loaded with current that it effectively 'drowns' in its own magnetic field and there is a sharp cut-off, as at X in Fig. 7.6, where it ceases to be a superconductor and 'goes normal', i.e. it behaves like an ordinary piece of metal wire. Less dramatic, generally, is the fact that Ohm's law for a piece of metal wire is not the whole truth about the resistance of metals to the flow of electric current, for the resistance of most metals rises with increase in temperature. Thus the passage of current produces heat losses that raise the resistance, and the graph of current (result) against applied voltage (cause) is not the simple straight line of Ohm, but a curved line which rarely gets the opportunity to pass through a maximum, but is terminated by the wire melting, as in the 'blowing' of a fuse.

So we conclude that in matters of saturation there are, as usual, 'all shades of grey' that merge into each other and thwart our ability to 'explain' the whole in one simple argument.

* Superconductors are metals such as niobium-tin which, when cooled to about $-270\,°C$ (usually by immersion in liquid helium), lose all but a microscopic fraction of their electrical resistance. When a current flows in a superconductor, however, it still produces magnetic field as if it were an ordinary metal at room temperature. An excessive magnetic field of this kind can cause the metal to return to a 'normal', i.e. resistance-ful state.

Feedback

The term 'feedback' is a relatively modern one, having its roots in
the beginnings of control theory. The latter subject is inextricably
linked with the theory of amplifiers, and the recognition of an
amplifier came perhaps only with the invention of the triode valve,
although even its artificial forms (for example the action of the
crystal of the 'cat's whisker' in early radio sets) were known before
they were recognised as part of the powerful amplifier/feedback
concept. Certainly biologists were aware of the system as a whole,
for one of them coined the phrase: 'More cats, more clover',*
which exemplifies in a relatively simple manner the inter-
dependence of all living things in a system so complex that it
denies formulation by the most brilliant of human minds.

The idea is illustrated by the fact that cats catch mice and
mice (field mice) eat bees, both of which are *limiting* phenomena,
preventing the uncontrolled multiplication of any one species,
and seen in terms of the engineering amplifier as '*negative* feedback'. Add to this the fact that bees pollinate clover, a mutually
beneficial effect and therefore termed '*positive* feedback', and you
have the phrase explained, thus:

More cats means fewer mice	(negative feedback)
Fewer mice means more bees	(inversion)
More bees means more clover	(positive feedback)

The result of *more* cats is therefore *more* clover. Note how the
two negative terms cancel out, so far as propagation of the ultimate species (clover) is concerned. It is as if we represented all
negative systems as (-1) and all positive systems as $(+1)$ and used
algebraic multiplication to get the result. Thus the relationship
between cats and clover is $(-1) \times (-1) \times (+1) = +1$. Feedback
is as fundamental as that.

But the electrical engineer needs to *quantify* his algebra and it
is then that biology can benefit, if only a little, from the complex
artificial object. The benefit is only 'a little', because in order to

---

* Attributed to Charles Darwin.

get useful answers the characteristics of each stage of amplification need to be known (as we shall now see), and in the case of biological systems this is generally well-nigh impossible.

The basic electrical system for amplification and feedback is illustrated in Fig. 7.7a. A is an amplifier that can be anything from a rotating electric power generator such as we find in huge power stations to a mica chip a few millimetres square that carries a printed circuit containing six transistors and is itself a part of an elaborate computer. For our purpose all we need to know about it is that for an input voltage v it produces an output voltage G times that value (Gv). We say that the amplifier has a 'gain' of G. The power that is needed to sustain a far greater output than input comes from another source (shown as 'power source' in Fig. 7.7a) but it can be of any form (in the power station generator, for example, it is a steam turbine and the input power is mechanical).

The idea of feedback is introduced in Fig. 7.7b, where a fraction $1/k$ of the output is *fed-back* to add to, or subtract from, the existing input. If it adds, the feedback is positive, if it subtracts, it is negative. The 'box of tricks' labelled '$1/k$' is the unit that decides whether the sign is plus or minus, whether the value of k is constant and whether the feedback is instantaneous, or has a time delay imposed upon it. As might be expected on 'common sense' grounds, positive feedback leads to *greater* overall gain, whilst negative feedback leads to *smaller* gain.

But notice the deadly ingredient inherent in positive feedback. If the value of k should equal the value of G, the overall gain goes to infinity. The system has exploded.* In such situations some other factor not considered in the feedback system always comes to the rescue, for infinity is an unrealisable extreme. 'Saturation' is one of the most common of these other factors. In the case of human population, it is usually (up to now) lack of food that finally restores stability, but the experience of this kind of corrective factor is horrific.

---

* Not unlike the present trend in world population where improved medical knowledge has enabled a greater percentage of babies to reach maturity and where a greater number of mature humans will, on average, produce a greater number of medical workers.

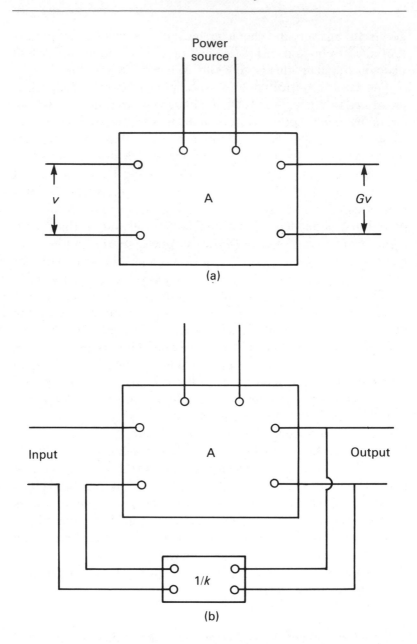

Fig. 7.7. Simple amplifier theory. (a) An indication of power levels. (b) The incorporation of feedback.

Where there is a time lag in the $(1/k)$ box the system will often take on instability of a different kind. Instead of trying to explode to infinity, the output will oscillate continuously between a maximum and a minimum value. A simple example will suffice to illustrate this behaviour. The principal enemy of a particular species of caterpillar is a particular species of fly (let us assume). The adult fly lays its eggs under the skin of the caterpillars, as described on p. 180, and when these eggs hatch, the fly grubs eat the caterpillar. Let us assume that in the summer of 1975, which was hot and sunny and conducive to much mating and egg-laying among butterflies, 100 000 000 female eggs of a particular butterfly species were laid by a greater-than-average adult butterfly population (1 000 000 females), which had been attacked in its younger days by a less-than-average number of adult parasitic flies. The eggs stay as eggs throughout the winter (the time lag). In the spring of 1976 the rare parasitic flies have a bonanza – there are 100 caterpillars per square metre on patches of ground containing the food plant, so 999 caterpillars in every 1000 get injected with fly eggs and perish. Only 100 000 females remain to lay that year, so at the 100 eggs per female that we assumed earlier, only 10 million eggs are laid but there are millions of parasitic fly pupae, and in 1977 the adult flies from these go to work on the mere 10 million caterpillars with a vengeance and reduce them by 99 990 in every 100 000, i.e. only 1000 female butterflies mature. The 'responsible' newspapers publish letters about 'the tragedy of the disappearing English butterfly', collectors are snarled upon and conservationists hold a flag-day to raise funds!

What actually happened in 1977 was that too many fly eggs were laid in the same caterpillar (on average) and there was not enough live caterpillar to go round. In fact 99 in every 100 fly larvae starved to death. The result was that in 1978 an 'average' number of parasitic flies tried to find homes for their young among a mere 100 000 caterpillars which were now hard to find, and only 1 caterpillar in 10 was attacked. So 10 000 females emerged to lay 100 eggs each and we were back to our million butterflies, and increasing. As for conservation and letters to the newspapers, no-one expressed concern for the 'endangered' species of parasitic fly in 1976, but in 1979 the glut of butterflies decimated a particular garden vegetable and there were letters on the theme: 'Science ought to be able to control these pests.'

This account is, of course, exaggerated and grossly over-simplified, merely to illustrate the mechanism of feedback with phase lag, producing oscillation. (Contrast this with the *permanent* decimation of butterfly and parasite due to removal of 90% of the foodplant, as described on p. 180.) Surely then, the opposite to phase lag should have powerful stabilising effect? Well the opposite of 'lag' is 'lead' and this implies at first sight a need for clairvoyance in order to implement it! But this is not so. All that is needed is a knowledge of the natural frequency of the oscilla-tions that are likely to occur. 'Phase advance' is a well-used tech-nique in man-made control systems. Nature, of course, was not slow to develop it, and again a living example will serve us best in appreciating the mechanism. Bats in flight prey upon night-flying moths, scooping them up with their shoulders if they 'just miss' with their mouths directly. But their detection mechanism is not 'vision' like our own, using light waves at a speed of 300 000 km/ second (186 000 miles a second), but acoustic waves travelling at a mere 0.32 km ($\frac{1}{5}$ mile) a second. Even so, the bat has to 'wait', albeit a mere $\frac{1}{20}$ second, for the echo from, say, 15 m (50 feet) but it has *given away its presence*, provided the moth can detect the acoustic signals, and many of them can. So the moth has the bat's flying time for 15 m (50 feet) to take evasive action – the phase advance that it needed to survive.

The phenomenon of feedback can be seen to be more funda-mental than some others described here, notably explosions and waves (oscillations). It is a vital factor in the struggle for survival, even against an unknown factor such as the appearance of humans on earth.

Resonance

This word obviously has its origins in the world of sound, but the phenomenon that it describes occurs in almost every facet of science and life. It can be regarded in its most fundamental sense as occurring when two things *fit*, like a peg in a hole, a plug in a socket or a key in a lock. Perhaps the most sophisticated way of describing it is as the supreme combination of positive feedback and saturation. Of these two descriptions the first is over-simplified, the second is too complex. Somewhere between lie the more commonly found examples, so let us begin with the simple example shown in Fig. 7.8. A heavy weight, say 1000 kg (1 ton), hangs by a stout rope. Attached to the weight is a fine thread of unspun silk (breaking strength less than 100 g or 3.5 oz). The problem is to cause the weight to swing as a pendulum through an angle of ±10°, only by pulling on the silk. At first this seems impossible, but if the *'natural' frequency* of the

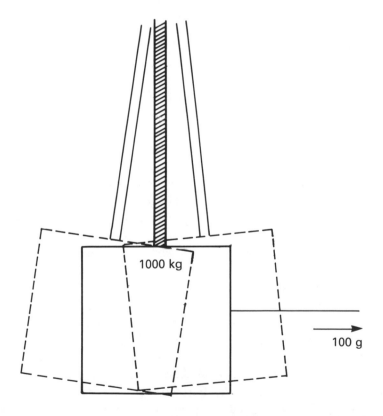

Fig. 7.8. Swinging a heavy weight by means of a weak thread.

weight and its rope, as a pendulum, is known, then the problem can be solved by pulling the thread sideways for only, perhaps, 10% of the time of one swing, but repeating the pull once per natural period of swing. After a few thousand such pulls the weight will have achieved its overall 20° of arc. You have, so to speak, 'tuned-in' to the natural frequency of the pendulum; you have *resonated* with it.

The idea itself is simple: you only reinforce, once per cycle, the motion or other result that would occur naturally were the system to have attained the necessary energy from somewhere else. The effect has been known for centuries in military establishments for marching soldiers have long been told to 'break step' when crossing a bridge, in case the natural cadence should coincide with the natural oscillation period of the bridge and the resonant effect should smash the bridge. The phenomenon of resonance could even be described as the reinforcing of a pencil line on a piece of paper by going over that same line again and again to make it stand out.

In the more sophisticated technologies, radio transmission for example, an aerial works best when its length is an exact fraction of the wavelength of the radiant energy. If you live near to a TV transmitter, almost any odd piece of wire will do as aerial, because the signal is so strong. But in the fringe areas, or where large buildings or trees or hills strain the resources of the transmitter, the design of the aerial must 'match' (and this word, now much used in technology, is almost synonymous with the words 'in resonance') the wavelength of the radiation involved. In a typical TV aerial the aerial proper is backed by a reflecting stub and preceded by a series of quarter-wavelength 'matching stubs' to concentrate the signal on to the receiving rods.

In the development of radar during World War II, the number of stubs used was often large. The mounting rod was often curved as here, and it occurred to me in those days that if an array such as this were to be turned on its side it would resemble the antennae of certain moths, well-known for their ability to communicate between male and female over very long distances. I have seen 12 miles quoted as a possible communication distance, but few scientists accept this. One mile, however, is generally accepted as a possible range for moths, although the mechanism is invariably said to be that of smell. Only a few moments' thought will convince most engineers that the sense of direction obtainable from an olfactory mechanism could never achieve the required accuracy at such distances, and in 1943 I wondered whether the word 'antennae' was chosen for radar aerials because they looked like the antennae of moths. Thus I embarked on a little private research of my own and found that the possibility of the moths' communication via their antennae might indeed be an ultra-short wave electromagnetic phenomenon.[1]

Some 20 years later, the effect was confirmed by a professional entomologist, Dr Philip Callahan,[2] who published widely and is still not yet believed by the majority of entomological scientists. Nature uses resonance, of course. The acoustic communication, at ultrasonic frequencies, of the mosquitoes brought about their downfall in Panama, once the nature of the phenomenon had been established.

Natural shapes – the golden ratio

Of the three groups of shapes I have so far elected to describe (based on the straight line and right-angle, the circle and the helix), the last is probably the one with greatest involvement in natural shapes, for it embodies *growth*.

At this point it is appropriate to introduce a third fundamental number, to add to e and π. Generally known as the 'golden ratio' and denoted by the Greek letter ϕ, it can be expressed in terms of a rectangle of such proportions that the ratio of the sum of two adjacent sides to the longer side is equal to the ratio of the longer side to the shorter.

It can easily be shown to be equal to $(1 + \sqrt{5})/2$, $= 1.6180339$ ... which does not, on the face of it, seem very exciting, but its geometrical and numerical significance is very profound. Figure 7.9 shows a number of rectangles and you are required to say which one you would choose to illustrate to the uninitiated a 'typical' rectangle. Most people choose (d), which is the one with

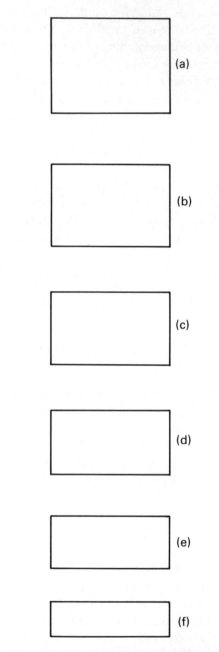

Fig. 7.9. (a) to (f) Which of these is a typical rectangle?

sides in the golden ratio. Artists throughout hundreds of years have known that this ratio is 'pleasing' and have used the dimension to proportion their pictures.

If a rectangle be drawn, having its sides in the golden ratio, the result of drawing a square inside it, using the shorter side as one side of the square, as shown in Fig. 7.10a, will be to leave a

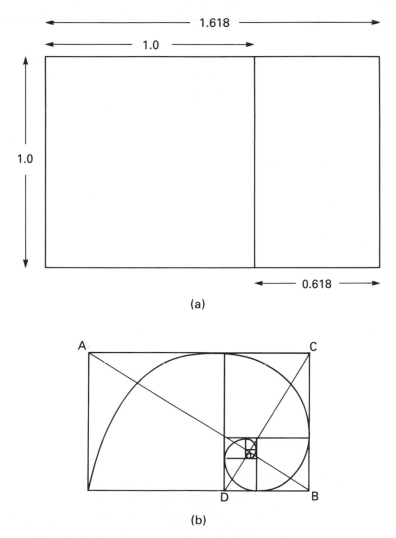

(a)

(b)

Fig. 7.10. Development of a helix from golden rectangles.

rectangle with sides in the same ratio, since $1/0.618 = 1.618$.*

This process can therefore be repeated *ad infinitum*, as shown in Fig. 7.10b, and a helix can be drawn joining corresponding corners in each rectangle. The significance of this particular helix is that if its scale is increased by *any* finite number of times, its *shape* will never change, i.e. if you look at a tiny, central portion of the helix through a microscope, you will appear to be looking at the original helix unchanged. It is therefore a natural curve for anything that is required *to grow and to retain its topological identity*. In modern terms, it is a simple 'fractal'.

The helix formed by the golden ratio rectangles is a *logarithmic* spiral insofar as its centre can be located as the intersection of any pair of diagonals from successive rectangles, as shown at AB, CD in Fig. 7.10b, and any line drawn outwards from the centre will cut the helix at distances that form a logarithmic series, i.e. the logarithms of the distances are an arithmetic progression. Thus the helix links ϕ with e, since the latter is the base of natural logarithms. Incidentally, the ratio of any two diagonals such as AB/CD in Fig. 7.10b is itself the golden ratio.

Numerically, let us first consider a series of fractions such that the top of each new fraction is equal to the sum of top and bottom of the previous fraction, and the bottom of each new fraction is equal to the top of the previous fraction. Thus if we start at $^2/_1$, the next term is $(2 + 1)/2 = {}^3/_2$, and then $(3 + 2)/3 = {}^5/_3$, and so on. The series rapidly approaches fractions that constitute the golden ratio, thus:

$$\frac{2}{1}, \frac{3}{2}, \frac{5}{3}, \frac{8}{5}, \frac{13}{8}, \frac{21}{13}, \frac{34}{21}, \frac{55}{34}, \frac{89}{55}, \ldots \qquad (7.1)$$

These fractions, expressed as decimals, are respectively:

2.00, 1.50, 1.$\underline{6}$ 1.625, 1.$\underline{615384}$,

1.$\underline{619047}$, 1.61764705882352941, 1.6$\underline{18}$, . . .

(The underlining implies that the system of digits so underlined repeats thereafter for that particular number.)

Thus a mere nine terms are needed to produce the first four

---

* Or more exactly $(1 + \sqrt{5})/2 - 1 = (\sqrt{5} - 1)/2 = 2/(\sqrt{5} + 1)$.

digits correctly. But the starting numbers need not be as simple as are 1 and 2. We could start anywhere, for example:

$$\frac{48}{3}, \frac{51}{48}, \frac{90}{51}, \frac{150}{99}, \frac{249}{150}, \frac{399}{249}, \frac{648}{399}, \frac{1047}{648}, \frac{1695}{1047}, \ldots \quad (7.2)$$

yields 1.6189 for nine terms, which is again correct for the first three places. This result is less surprising in the context of the spiral being viewed through a microscope. Growth can begin at any stage, otherwise there would be only one species of living creature.

The fundamental nature of the golden ratio is demonstrated by its ability to be expressed numerically solely in terms of the number 1 – and this in more than one way. It could be argued that just as an atom is taken to be *real*, whilst magnetic field is a concept of the mind, so the number 1 (the 'atomic structure' of *all* numbers) is real and processes such as 'division', 'reciprocal', 'square root', 'negative', etc. are purely concepts. There is no difficulty in writing down $\sqrt{-1}$, nor is there in seeing at once that it is never expressible in terms of 'ordinary' numbers, but as a concept it is extremely useful.

The golden ratio, combining only the number 1 with concepts, is given either by:

$$\phi = 1 + \cfrac{1}{1 + \cfrac{1}{1 + \cfrac{1}{1 + \cfrac{1}{1 + 1 \ldots \text{ to infinity}}}}}, \text{ or}$$

$$\phi = \sqrt{1 + \sqrt{1 + \sqrt{1 + \sqrt{1 + \sqrt{1 + \ldots}}}}} \text{ to infinity}$$

Compare these expressions with the series for e, written in terms of 1 and of concepts:

$$e = 1 + \frac{1}{1} + \frac{1}{1(1+1)} + \frac{1}{1(1+1)(1++1)} + \frac{1}{1(1+1)(1++1)(1+1+1+1)}$$
$$+ \ldots \text{ to infinity}$$

and the other concept and formula:

$$e^{\pi\sqrt{-1}} = -1 \quad (7.3)$$

and one has a feeling that it ought to be possible to express π in terms of 1 and of fairly simple concepts, but so far as I know, this has never been done.

A series of numbers such as 3, 48, 51, 99, 150, 249 ..., in which the first two may be chosen randomly but thereafter each number is the sum of the previous two, is known as a Fibonacci series, although originally the name applied only to the series 0, 1, 1, 2, 3, 5, 8, 13, 21, 34, 55, 89, 144 ... Fibonacci was a thirteenth century mathematician, known as 'Leonardo of Piza', who apparently, just for fun, toyed with series, one of which contained a summation series as numerators and a 2^n series as denominators, thus, 0, 1, $\frac{1}{2}$, $\frac{2}{4}$, $\frac{3}{8}$, $\frac{5}{16}$, $\frac{8}{32}$, $\frac{13}{64}$, $\frac{21}{128}$ ... But it had to wait some six and a half centuries before it was recognised as relating to the natural formula for growth. It will now be apparent by turning back a few pages that the series (7.1) and (7.2) are generated directly from Fibonacci series. It is the prime link between the two concepts. At the same time, another property of the golden ratio (as if it needed one!) is that if it is used in an additive series beginning with 1 and x, thus:

$$1,\ x,\ (1 + x),\ (1 + 2x),\ (2 + 3x),\ (3 + 5x),\ (5 + 8x) \ldots$$

the ratio of any two consecutive terms is equal to the golden ratio if x is itself the golden ratio (ϕ). Hence the series:

$$1,\ \phi,\ (1 + \phi),\ (1 + 2\phi),\ (2 + 3\phi),\ (3 + 5\phi) \ldots$$

is identical to the series:

$$1,\ \phi,\ \phi^2,\ \phi^3,\ \phi^4,\ \phi^5 \ldots$$

and ϕ is the only number for which this is true.

There is a society bearing his name that publishes a quarterly journal. Such is the mass of unravelled material in numerology that one could compare it with the amount of data available from living structures, of which *Gray's Anatomy*,[*] now occupying 1471 pages, is only concerned with the structure of one species! One might even have said that the two were interdependent, with a fair degree of likelihood of being correct, were it not for the fact

* See Reference 1, Chapter 1.

that less complex living structures are more clearly connected with the Fibonacci numbers. What is more, they show the inevitability of the connection in all things that grow.

Helices in plants

For the better appreciation of Stevens' work, it is necessary to be familiar with the mathematical and geometric structure, of the planar helices in particular. Fig. 7.11a shows a double-entry helix which has a single point of inflection at its centre. Such helices can be Archimedean, logarithmic or otherwise. It is interesting to note in passing that the ancient Yin–Yang symbol of the Chinese mystics was a double-entry helix reduced to its simplest form. Readers should note that a double-entry helix, continued outwards to infinity, divides the whole of space into two equal parts. No line can be drawn between the coils (like the beginning of such a line, shown dotted, in Fig. 7.11a) without meeting a dead end at the centre. Furthermore, all multiple-start helices of higher order than two must converge on a point at the centre, unlike the double-start helix. Figure 7.11b shows an example, pointing out that whilst combinations of these helices will depict precisely the layout of celery stalks, pine cone florets, daisies, sunflowers, etc., the plants themselves use no mathematics. They simply grow the stalks where there is most room and *we* introduce the mathematics just to describe three-dimensional growth on a two-dimensional picture.

Let us draw a single helix (Fig. 7.12a) and a double helix (Fig. 7.12b), by marking off along rectangular co-ordinate axes the consecutive distances $-1, 2, 3, -4, -5$ etc. and joining them for Fig. 7.12a. For Fig. 7.12b, plot points $3, -3, 6, -6, 9, -9$, etc. and use these to draw the double-entry helix. Now superimpose the two as shown in Fig. 7.12c. Then shade in and number the areas trapped between the helices (it is best to leave a border around each) as shown in Fig. 7.12d. This is precisely what you see when you take a cross-section through a celery plant near to its base.

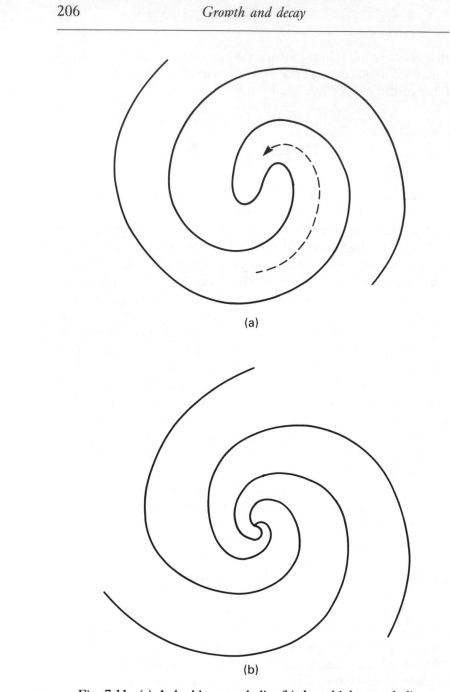

Fig. 7.11. (a) A double-entry helix. (b) A multiple-start helix.

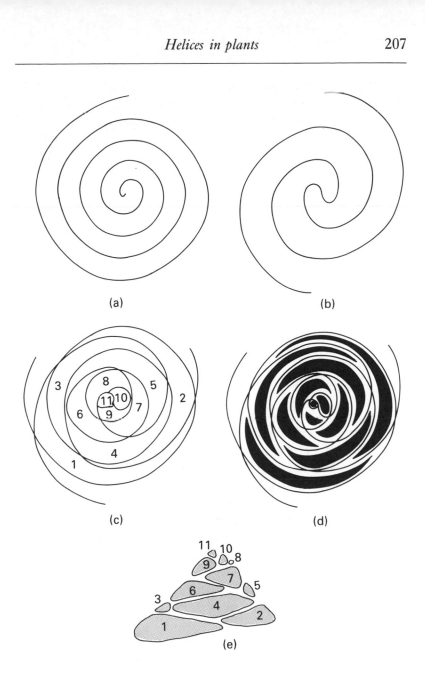

Fig. 7.12. Modelling a plant from helices. (a) A single-start, left-handed helix. (b) A double-start, right-handed helix. (c) Superimpose (a) and (b). (d) Spaces in the pattern are highlighted. (e) Side view of an ascending helix.

Nature's reasons for doing this are easiest to explain by means of an example, such as the celery. The helical pattern arises because the leaves develop one at a time (and, of course, not all plants are restricted to such development). Each leaf or stalk grows in the largest available gap between its neighbours (see Fig. 7.12d). Less obvious is the fact that any new stalk slides over the older pair ahead of it, as it grows, so that its centre is closer to the centre of the outermost, or older, of the two previous stalks than it is to that of the younger. The stalks attach themselves to the stem (actually to a base stem known as a 'meristem') so that an ascending helix viewed from the side appears as shown in Fig. 7.12e (after Stevens) for the celery. Notice how stalk 4 sits low between 1 and 2, thus leaving room at the sides for stalks 6 and 7 to grow between 3 and 5. In effect, the stalks in one layer squeeze through the gaps between stalks in an outer layer to touch stalks of a still further outer layer, a process that results from compression of the diagrams which is to be found in the younger parts of the plant.

Less compacted leaves which are well spaced around a plant stem exhibit the phenomenon that the numbers of the leaves that appear directly above each other are part of a Fibonacci series. It has no significance, so far as we know, to the function of the leaves. It was how they were 'born' that set the pattern. Northrop points out that similar arrangements occur in pine cones, flower petals, the heads of lettuce and the layers of an onion.[3] Stevens points out that just as the celery has a 'phyllotaxis' of $\frac{1}{2}$ (single and double helices), a typical pine cone and pineapple both have 8 laid on 13. Other pine cones can have the $\frac{2}{3}$, $\frac{3}{5}$ or $\frac{5}{8}$ super-positions; I have seen a daisy with 21 on 34, whilst sunflowers are also commonly found with fractions $\frac{55}{89}$ and $\frac{89}{144}$ and one of $\frac{144}{233}$ has been recorded.

But let us never lose sight of Stevens' summing up: 'The plant is not in love with the Fibonacci series; it does not seek beauty through the use of the golden section; it does not even count its stalks; it just puts out stalks where they will have the most room. All the beauty and all the mathematics are the natural by-products of a simple system of growth interacting with its spatial environment.' One can only gasp and perhaps add: so *that's* why the Trojan war was fought over Helen!

The growth of a Man-made object

Among the articles published by the *International Federation of Medicine Student Associations*[4] is one by Richard Moore. The subject is the patterns in ancient mosaics. Without going deeply into the subject here, it will suffice to say that ample evidence is available to deduce that the mosaic constructors, working with randomly broken pieces of flat tile within a given size range, worked from the centre of a floor outwards, always choosing the next piece to be laid at random and placing it in such a position that it touched at least three tiles already laid. There was no question of pre-selection of a piece 'to make it fit'. There was no attempt to rectify a 'mistake' in what appeared to be regular patterns. This finding was a most important result for it suggested to Moore that the patterns might be the *inevitable* result of purely random packing of 2-D shapes in a 2-D space. In other words, 'order', as we normally define it, comes out of chaos **of necessity** and not by chance!

Moore made over 300 000 measurements of the lengths of patterns in Roman and Greek mosaics in Italy, North Africa, the Eastern Mediterranean and the UK, and found that the same measurements were turning up far more frequently than would be predicted by ordinary probability theory. The mosaics in question were located hundreds of miles apart across the earth and were completed at periods spanning a thousand years. For example, 6000 of the measurements were 9.6 cm. By 1964 Moore had examined over 1000 mosaics[5] and found concentrations of distances to such an extent that (to the nearest millimetre) 89% of the third of a million measurements were each one of the series:

1.2, 2.4, 3.6, 6.0, 9.6, 15.6, <u>21.6</u>, 25.1, 40.7, 65.8, 106.5, 172.4, 278.9, 451.3, 730.2, 1181.4 (all measurements in cm).

This is so clearly a Fibonacci series, the only exception being the 21.6 which is the sum of the previous number and the next but *two* previous, i.e. 6.0 and 15.6. Of the remaining 11%, 9.5% were simply multiples or sub-multiples of these distances. All but 1.5% fitted the series.

Moore formulated a general theorem, thus: 'For any system composed of discrete particles (the latter being of, between limits, randomly dissimilar sizes and shapes) there exists a set of absolute values at which symmetrical phenomena within the system will tend to occur.'

One of the most readily available examples with which to test the theorem is the spacing of words in printed books and papers. Take any page of a book at random and de-focus your eyes by looking as it were 'through' the book to a point beyond. You will almost always see vertical white lines running down the page where the ends of words have lined up with each other. If such lines occur mostly in the right-hand half of each page or column, it is probable that the subject matter of the book is popular science, or it may be a novel. If the lines occur in the left hand half, the book is almost always a child's story book. But if most pages contain no clearly defined lines at all, it is likely to be a highly technical book or a paper for a learned society.

The 'secret' lies in the phrase 'within limits', for the distance of the lines from the left-hand margin depends on the average length of the words and the range of their lengths. In 1966 Moore concluded: 'This aspect is at present being examined in a fundamental way, for it is possible that the packing hypothesis is a special case of a concept suggested here, that random disorder does not exist in reality.'[6]

The axe had been laid to the foot of the tree called 'entropy'.

By 1972, Moore had shown mathematically 'that a maximum to the degree of disorder exists' when particles of randomly varied size and shapes within limits are aggregated, insisting powerfully: 'This is not an end-effect, the same is true for an infinite population.'[7] He goes on to show that the positions of the greatest misalignment (randomness) are just as organised as those of least, and he adds: 'Order and disorder here appear to be interchangeable', and indeed one of his publications bore the title 'Order as an apparently spontaneous product of chaos.'[8] He pointed out that the rules for packing mosaics were the same for bubble rafts, sand crusts, scums and the cells of leaves. He bridged the animate and the inanimate such as had never been attempted previously. It is interesting also in this context that the late Alan Turing, the mathematical genius who unravelled the German 'Enigma Code' in World War II, spent the last summer of his all-too-short life with a map measurer and a planimeter, recording masses of data relating the perimeters to the areas of tens of thousands of leaves.

8

Nature – master technologist

On pp. 135–136 I drew attention to Gilbert Walton's comparison between the animal and the machine in respect of bilateral symmetry.

There are exceptions, of course. The flat fish that lie on the seabed are still desperately trying to restore symmetry for the eyes by bringing one from underneath, thus re-pairing it with the one already on top, and they have got it about as far as the edge that divides top and bottom. Martin Gardner himself pointed out this example and others, including the fiddler crab with its disproportionately sized claws, and the wry-billed plover from New Zealand, which uses its right-twisted bill to turn over stones, nearly always from the right. The cross-bill, on the other hand, has its beak crossed in either direction and reminds us of the legend that its bill became twisted as it vainly tried to pull the nails from the cross when Jesus was crucified. Its red breast became stained with blood in the attempt.

In the same article Gardner described some *very* strange creatures, displaying asymmetry in quite another manner. The hagfish is like an eel, but it can tie its long body into a knot, of either handedness, and by sliding the knot from tail to head it scrapes the slime from its body. The self-knotting technique has a dual purpose. (One begins to wonder what organ in nature does not!) It uses it to obtain leverage when tearing food from a large, dead fish. A third use is to escape from an enemy. (A second knot-maker has already been discussed in Chapter 4, p. 100. A full reference is given in the list of references.[1])

Religion and evolution

One cannot elaborate on the marvels of living things, nor speculate on the origin of it all for very long, before religion enters the argument. The number of Christians who reject the Theory of Evolution is decreasing, mostly I suspect because it is not incompatible with the idea of a Creator. It simply means that one has to go further back in the chain to meet the need for creation. That we are not the only intelligent life in the universe is tantamount to being self-evident. That we are the oldest is much longer odds than winning a million pounds for a penny on a football pool. Give humans the *same* rate of scientific and technological progress as that of the last 50 years for another thousands years and where might we be by then? But give us another million years? We could have evolved into pure energy (or thought). We might have become 'Gods' for lesser-developed species in our own galaxy. By inference, therefore, the likelihood that 'God created humans in his own image' is negligible. Martin Gardner wrote, 'A god whose creation is so imperfect that he must be continually adjusting it to make it work properly seems to me a god of relatively low order, hardly worthy of worship.'[2]

Evolution is a process of trial and error. It is as if an operator were using a copying machine to reproduce a page from a book of poems and a part of a word was blurred by a defect in the paper. A reader thought it said 'emotion' whereas the original had been 'potion'. The reader used the quotation in a book he was writing. The book became famous, the original book faded in popularity and decades later the majority of mankind never knew the original word. This happens more frequently in transla-

* A complete article on the hagfish, written by David Jensen, is to be found in *Scientific American*, February, 1966.

tions than in straight copying.* *The Bible* must contain many sentences whose meanings have changed. One needs only to read the *New English Bible* to see just *how* much has been lost in the 'modern' (dull) language.

But the photocopier error is a much better analogy to the evolution process. In the first instance, the accuracy of a photocopier approaches that of the genetic code when DNA molecules 'unscrew'. Secondly, only if the new word makes an *improvement* does the 'mutant' version survive to overthrow its ancestor. But by the time we have taken as many photocopies as nature has produced babies (of all species)† even Shakespeare might have

---

* A well-known firm that carries out much business with the Spanish-speaking countries of South America always has its Englishman-composed letters translated into Spanish and then re-translated into English by a different interpreter. On one famous occasion, the phrase 'It is respectfully brought to your attention that . . .' came back, after double translation, as 'Get this . . .'!

† If one in ten of the world's population makes 100 photocopies each per day (a generous estimate) and each creature that is alive on earth today (and can be seen without using a lens or microscope) produces only one egg, seed or equivalent that survives to propagate the species (the figure that alone ensures stability for the ecology), then it will take us a million years of photocopying to produce as many replicas as nature provides in a day!

This is based on the following estimates. (Readers finding serious errors in my estimates can substitute their own figures without destroying the message.)

The locust swarm that was recorded in 1889, crossing the Red Sea, was estimated to contain about 250 000 000 000 insects (weighing about half a million tons): (2.5×10^{11})

Allow only 40 times this figure for the greatest number of insects of any species. Total insects of that species: (10^{13})

Assume the average insect species to have a population of only 1/1000 of this (sheer guess). Number of insects per species, on average, is then: (10^{10})

There are over 400 000 species of beetle, nearly 200 000 of butterfly or moth. Allow a million species of insect. Number of insects now alive: (10^{16})

Allow each species 1 year to reach maturity. The number of reproductions annually is therefore also: (10^{16})

The world's population is about 4.5 thousand million, say: (5×10^9)

Assuming that 1 in 10 makes 100 photocopies a day, the total photocopies a year is of the order of $5 \times 10^9 \times 10 \times 365$, or about: (2×10^{13})

It will take 500 years to compete with the insects alone.

It has been estimated elsewhere that of the 10^{33} living creatures on earth, 75% are viruses. But even the remainder, 2.5×10^{32}, is almost unimaginably large. It would take over 1 000 000 000 years of photocopying at the specified rate to produce as much new literature by mutation as the earth has produced species (allowing an average of 50 000 words per book and only one word in every million mutations to be seen as an 'improvement').

been improved upon to an unbelievable degree. Thirdly, the word 'improved' relates to the mind of the reader, in the case of the photocopier, and the mind is more likely to cause the changes than the misprints are to succeed in their own right. In the case of evolution, the 'mind' is replaced by the 'environment'. It was the Ice Age that demolished species by the million. It has been estimated that of all life species, other than viruses, that have ever lived on earth, over 99% are now extinct.

So much for conservationism!

I cannot leave the subject of evolution without reference to a fascinating article by an ex-colleague from Manchester, Professor H. S. Lipson.[3] Not only does he place a question mark against evolution, but also against Lamarckism, about which I must comment also.

> I have always been slightly suspicious of the theory of evolution because of its ability to account for *any* property of living beings (the long neck of the giraffe, for example). I have therefore tried to see whether biological discoveries over the last thirty years or so fit in with Darwin's theory. I do not think that they do.
>
> In the last thirty years we have learned a great deal about life processes (still a minute part of what there is to know!) and it seems to me to be only fair to see how the theory of evolution accommodates the new evidence. This is what we should demand of a purely physical theory . . .
>
> I shall take only one example – breathing . . . We know that it involves a complicated chemical, haemoglobin, the molecule of which contains several thousand atoms. Thanks to the work of people such as Perutz and Kendrew, we now know how the molecule is constructed and we know the conditions under which the oxygen molecules are held and released. We still do not understand the nature of the forces; they are necessarily very delicate, and for this reason a large molecule is needed.
>
> Darwin says 'If it could be demonstrated that any complex organ existed which could not possibly have been formed by numerous, successive, slight modifications my theory would

absolutely break down.' I know that haemoglobin is not an organ but the principle is the same; I do not see how the haemoglobin molecule could have evolved . . .

There is another theory, now quite out of favour, which is based upon the ideas of Lamarck: that if an organism needs an improvement it will develop it, and transmit it to its progeny. I think, however, that we must go further than this and admit that the only acceptable explanation is *creation*. I know that this is anathema to physicists, as indeed it is to me, but we must not reject a theory that we do not like if the experimental evidence supports it . . .

According to Darwin, when Newton put forward his theory of gravitation, Leibnitz accused him of introducing 'occult qualities and miracles into philosophy'. What was this gravitation? How could two inanimate bodies attract each other? Newton replied laconically 'Hypothesis non fingo'. When I am asked to describe my ideas of the Creator I also say 'Hypotheses non fingo'!

Darwin was also fond of the quotation 'Natura non facit saltum' (Nature does not make jumps). I wonder what he would have thought of the quantum theory!

In a BBC TV programme in 1981, Eldridge remarked that 'Paleontologists, pretending to believe in gradual change, have known for over a hundred years that it was not there.'

On the subject of creation, one is reminded forcibly again of D'Arcy Thompson:

Time out of mind it has been by way of the 'final cause', by the teleological concept of end, of purpose or of design in one of its many forms (for its moods are many) that men have been chiefly wont to explain the phenomena of the living world; and it will be so while men have eyes to see and ears to hear withal. With Galen, as with Aristotle, it was the physician's way; with John Kay, as with Aristotle, it was the naturalist's way; with Kant, as with Aristotle it was the philosopher's way. It was the old Hebrew way, and has its splendid setting in the story that God made 'every plant of the field before it was in the earth, and every herb of the field before it grew.'

Even if one does not go quite that far, the most materialistic scientists are agreed that Trilobyte fossils reveal that they evolved in short bursts of rapid change, that 96% of all species became extinct at the end of the Cretaceous period, and that there was a sudden explosion of new creatures for which we have no explanation. Some have argued that impacts of large meteorites initiated the changes, in which case evolution was more a matter of luck than of genes.

Professor Lipson also had a tilt at a law of physics, perhaps that most sacred of them all – the Second Law of Thermodynamics. This latter is associated with the concept of entropy whereby a body always degenerates towards increasing entropy, i.e. minimum available energy. Thus it condemns all things to slide from order into disorder. I have long believed that living things did not conform to this law and I am delighted to have Professor Lipson say it for me.

> The beautiful and meticulous system which we call a living
> being is an ordered one; each atom must be in its right place.
> Generally systems tend to disorder – maximum entropy.
> Living beings seem to disobey this rule.
>   . . . if we were to regard the birth of an animal as regulated
> by the principles of thermodynamics, we must believe that
> the developing arrangement of atoms is that of lowest internal
> energy. My mind boggles!

A few years ago I wrote an article for a highly respected scientific journal but it was returned for modification with one paragraph in particular struck out with the words 'Pure Lamarckism' written in the margin. That a creature, having a skill (such as the solitary wasp that picks up a pebble in its jaws and uses it as a hammer to smooth the walls of its burrow, whilst its close relatives use their heads), should pass on an increased probability of its offspring doing likewise, seems to me not merely Lamarckism, it is common sense!

Only very recently, a new controversy arose in respect of Lamarck. Two scientists working in Canada claimed to have injected a mouse with the necessary drugs to allow a skin graft to be made on it, without its body 'rejecting' foreign matter. They claimed that its offspring were also able to take skin grafts. This should have re-established Lamarck without question (and without upsetting Darwin, I would have thought), but other workers were unable to repeat the experiments and the whole argument made news in the *Daily Telegraph* for 27 July 1981.

Learning, memory or the properties of matter?

No sooner had the first full-scale digital computer been launched on an unsuspecting world than those who wrote its first programs began to ask such questions as: 'Can the process be reversed?', i.e. 'If we fed it answers could it work out the questions?'. Still more profoundly: 'Can it *think*?' This last question was neither easily answered nor easily dropped and people began writing books on it; ideas on thought processes were revised. I think a lot of it stemmed from the invention of the 'random number generator'. The basis of this device was that in any piece of wire at any given time there are more of the free electrons moving in one direction than the other and that these constitute microscopic, but measurable, electric currents which, if sampled once per second, for example, will be the equivalent of tossing a coin for 'heads or tails' (since such machines work in binary numbers in which 1 and 0 are the only possible digits). But with this difference: there are so many million million free electrons in a bit of wire an inch long that the process is far more 'random' than tossing a coin (see also p. 163). After that, it was easy to program into the machine a faculty of judgement beginning with simple answers to questions such as: 'Is number X bigger than number Y?' and proceeding to much longer programs which ended with

---

* The failure of the human mind to appreciate true *randomness* was illustrated at this time by a firm which bought random numbers (yes, they are worth money!). They had been generated in binaries but translated into decimal digits and the purchaser claimed that the machine had malfunctioned because one of the numbers was 1906888805, and the block of four 'eights' was clear evidence of non-randomness. Had they been really unlucky they might have had a number with all 10 digits the same.

the question: 'Is machine A more efficient than machine B?' *Quality* could then be quantised and the machine could then build up case histories on any subject and begin to sit as judge, jury and executioner on problems hitherto seen as capable of solution only by a 'superior' human mind. It could be taught, as a child is taught, but more rapidly. Of course, the science fiction writers had a field day.

And all this occurred because the wretched machines had a 'memory'. The human memory is very complex. It takes account of most of the happenings of our earlier life and a sizeable amount of that of our ancestors, built into the genetic code in ways that are still obscure. But if proof be needed of the latter, how else are we to explain why an annual migration of Monarch butterflies from Canada, 3000 miles down the coast into California always collect on the same row of trees on arrival? These insects are, by our standards, pretty primitive creatures – and they never knew their parents!

We are obviously not going to unravel these questions in a couple of generations. Suppose we start from the other end of the problem where we *do* understand the rules of the game.

If you empty a bucket of lead shot on to a horizontal surface, it takes up a conical shape. In a classic demonstration during a lecture to the Society of Arts in 1883, the great engineer Osborne Reynolds showed that an identical shape could be obtained by pouring the shot into a rubber bag until it was full, then tying up the neck and dropping the bag on to the same horizontal surface.*

* He was defining the difference between rigid and non-rigid bodies. He filled a rubber bag with coloured water, tying a glass tube into the neck and filling it until the water level was clearly seen in the tube. Holding it by the neck, he asked a member of the audience to press on opposite sides of the bag. The water rose in the tube. He filled a second bag completely with lead shot. Then he poured in water to fill the spaces between the shot. Again the glass tube inserted in the neck had a water level about half-way up. He asked the volunteer to press the sides of the bag. The water level in the tube *fell*! During the pressure disturbance the shot had managed to re-arrange itself with more space economy.

A third bag was filled with shot only, well shaken and topped up and the neck tied. He punched the side of the bag and left a permanent imprint of his fist in the side. Then he re-opened the neck and filled up the spaces with water, shaking the bag a great deal (a) to let the shot attain maximum economy in

If soil is blown up from below the ground through a vertical tube, it too forms a cone. So volcanoes makes cones, molehills are conical, so are the entrances to the holes of earth bees. Size makes no difference. But we certainly don't credit the soil with having had a 'memory' and 'ancestors' in order to reproduce the same shape each time the same situation obtains.

I hope the point is emerging for you. Do the two ends of this argument really meet? Is the genetic code inevitable? If so, then the great religions of the world totter. But at least the main subject matter of this chapter requires no explanation. We can expect to see the same processes repeated in very different creatures and at the same time diversify by the random process, the mutations. Isaac Asimov, scientist and science fiction writer *par excellence*, expressed the same sentiments in a different way. He argued that if atoms of hydrogen and oxygen were allowed to combine at random to form molecules of only three atoms each, then they could be arranged as HHH, HHO, HOH, HOO, OHO and OOO. From such a mixture, 10 molecules are extracted at random. What are the chances of all 10 being HHO, which is *water*? Apparently 6^{10}, or $60\,466\,176 : 1$. But if the experiment is performed and hydrogen is burned in oxygen, *all* the resulting molecules are water, because water is the only three-atom combination chemically possible for oxygen and hydrogen, just as a cone is the only possible shape of heap that can result from friction-ridden, roughly spherical particles deposited in one spot. The rules that comprise the *mechanism* are different in each case. Asimov himself describes the rules of chemistry delightfully when he says that atoms are not like sticky marbles which, when shaken in a barrel, can stick together in any old way.

In the case of heaps of soil or whatever, the fundamental laws can be formulated in terms of the difference between *compressive* stresses, in which deformation occurs as the result of compacting of the actual particles of which the heap is composed, and *shearing*

---

packing, and (b) to remove all air bubbles. Finally he tied the neck below the water line. He then put the bag down on the bench and bounced a golf ball off it with great resilience. 'The bag, gentlemen,' he said, 'has become a rigid body.'

What a pity that the art of the superb demonstration to assist a lecture is a dying one.

stresses in which complete layers of substance are forced to slide over each other. In general, loose materials and to some extent *solid* materials are more ready to striate and form slip planes than they are to yield to compression, so new particles on top of the heap roll down until the angle of the cone is large enough for the friction to hold them above the particles below, and a block of concrete, or cast iron, placed under high compressive stress, fractures so as to leave a cone, for the same reason.

Martin Gardner, pointing out that the primordial conditions on earth may have made it difficult for amino acids *not* to form, continues:

> It is amusing to find so many well-meaning theists cringing
> back with horror these days at theories designed to bridge
> the gap between life and non-life by the operation of 'un-blind
> chance' – the union of chance and natural law.* It is amusing
> because it is easier to imagine *this* gap bridged than many of
> the later gaps in the history of life on earth. For example,
> chlorophyll had to be discovered,† as the means by which
> living units (plants) could use solar energy to manufacture
> starches and fats. Single-celled animals had to discover the
> short-cut of eating the plants.‡ Death and sex had to be
> invented by many-celled organisms capable of growing old
> and ceasing to function as a co-operative colony of cells.
> Animals had to discover how to eat other animals.

I have never seen the rules of life stated to succinctly, but if there is to be one, and only one, rule for all of it, then Pasteur wrote it, and Peter Stevens re-wrote it. Pasteur said: 'Life, as manifested to us, is a function of the asymmetry of the universe and of the consequences of this fact.' He went on: 'I can even

---

* The same sentiment was expressed by Sir William Bragg in the context of atomic energy. But we should note and marvel at the wisdom of someone who could predict the certainty of an event which remained uncertain for many decades before it became fact: 'A thousand years may pass before we can harness the atom, or tomorrow might see it in our hands. That is the peculiarity of physics – research and "accidental" discovery go hand in hand.'

† Discovered by primitive living things at the time, that is, not by biologists.

‡ From which the animal, as opposed to plant, kingdom never turned back.

imagine that all living species are primordially, in their structure, in their external forms, functions of cosmic asymmetry.'

Stevens pointed out that all evolution had taken place under the constraints of three-dimensional space (see also p. 147) ending with: 'The shape of space is a helix.'

So we are not to be surprised at what follows. It is a list, by no means comprehensive, of the many comparisons between civilisation's technology and that of nature. We may be disturbed, of course, by our failure to capitalise on copying the Grand Master, perhaps until now.

Textiles

Now that we can imitate natural fibres, we might begin by answering the question: why extrude nylon into thin strands, cut them into short lengths, spin them to make them stick together again and weave them to make continuous sheets from which to make clothes, when we could easily have produced them as continuous sheets in the first place? In the case of rainwear, of course, we do just that, but most clothes need holes for ventilation and, more important, a multitude of spaces in which low thermal conductivity air can be trapped to provide warmth. They need a flexibility beyond that of any continuous sheet and fashion demands other properties such as refusal to be creased, ability to be *permanently* creased (what fickle creatures we are!), ability to be washed or cleaned. They must be poor electrical insulators (to avoid becoming electrostatically charged), cheap to produce, durable; the list seems unending, but overall lies one glorious property without which few textile techniques would succeed. It involves, yet again, our old friend the helix. The theory of spun yarn was dealt with briefly in Chapter 6 (pp. 141–144).

Cotton fibres never attain individual lengths over 5 cm (2 inches). What then is the mechanism, known to the Ancients, that allows continuous cord to be made from this substance only, such that it will sustain high tension along its entire length? It is solely due to success in the art of making a large area of contact between one fibre and another so as to maximise the natural frictional forces present due to the nature of the surface of cotton fibres. In other words, it was micro-biology, exploited long before the microscope was invented. It had been discovered by experiment, by Faraday's way of doing science as opposed to Einstein's. It is pure engineering. An agglomerate of 5 cm (2 inches) fibres laid side-by-side and laterally compacted has virtually zero tensile strength longitudinally. But *twist* it, and the strength appears as if by magic. Each fibre is now a helix, and as tensile stress is applied, the fibres are forced together to increase the *pressure* between fibre and fibre so that natural frictional forces are multiplied by thousands and the surface area of contact becomes large as the empty spaces in the yarn are squeezed out by that same tensile strength that it is the aim to produce. Whether the discoverers of the mechanism looked at climbing plants first is unknown, but certainly many of these vegetables use the helical shape to obtain a bond between themselves and an existing structure. Failing this, they twist around themselves and then twist the resulting pairs to form ropes exactly in the way that we make ropes.

This introduces one further technique, that of the 'coiled coil'. We use it in the filament lamp to give it, as it were, a 'fourth dimension' of flexibility, to prolong its life as a substance that alternates between long periods in a state of white heat and long periods at room temperature. As with many things, it is the *changes* that produce the final rupture; it is the rapid expansions and contractions that produce metal 'fatigue'. Metals get tired, as muscles get tired, but rest is no cure for the metal. Compare then the coiled coil filament with the structure of a chromosome. Then compare the climbing plant with hand-made rope. The first art of textiles is the *spin*.

Weaving also uses contact area to give a virtually two-dimensional structural strength in any direction in its plane. The tensile strength was there in the yarn. Weaving binds two sets of parallel threads orthogonally to use components of both tensile stresses to resist any obliquely applied forces. It requires only a little imagination to see the connection between woven cloth, gyroscopes and electromagnetic devices.

Nature's spinners and weavers are common. The silk makers (spiders and caterpillars) exude the liquid that rapidly solidifies from 'spinnerets'. Their product still provides finer and more expensive garments for humans than any artificial fibres. Naturally crease-resistant, as thin as anyone would ever need for making fabric and with a natural tensile strength that is more than three times that of steel, it is the ideal substance for spinning and weaving. But the spider uses a less complex mechanism for building a web – spot welding is our nearest approach. The arachnid can produce sticky thread or smooth thread at will. The radial strands of a web are smooth. The circumferential threads are sticky and adhere naturally to the radial frame and to the insects it intends to trap. The spider's feet are coated with oil to prevent adhesion. The spider also tastes through its feet – the dual purpose implement yet again.

One of our modern techniques in binding surfaces together is to make a sticky substance which will not harden until it is mixed with a second substance. A popular product of this nature is marketed under the name Araldite. No doubt there are examples of this in nature which I have so far failed to locate, but one interesting phenomenon turned up whilst examining the many and varied abilities of spiders. One might call it 'anti-Araldite'. As a web is being spun, the silk emerging from the spinnerets would harden too rapidly (the surface/volume ratio decreasing in proportion to sectional radius of thread). From a separate gland, another secretion is poured over the thread as it emerges to allow the circumferential strands to remain sticky.

It has been estimated that there are some two million million spiders in Britain and that in a year they eat a weight of insects greater than the weight of the human population of the same country. Some of the huge tropical species can live for 25 years and can exist without food for up to a year. The arachnids are a force to be reckoned with.*

---

* Superstition has long had a place for spiders, 'bad luck to kill a spider' being the comment. In 1760 a Dr Watson wrote of a fever cure that involved swallowing a spider, 'gently bruised and wrapped in a raisin or spread on bread and butter'. Many people must have preferred the fever!

A spirit of adventure

We rightly admire the courage of those who venture into space. Space explorers are the heroes of most of science fiction. But we are not lone adventurers in this world. As with several facets of life involving ability, the arachnids come out near the top of the 'league table'.

One day I was sitting on a garden seat when I saw a silken thread stretching from the back of my chair to a low wall about three feet away. The thread glistened in the sun so that for a moment I failed to notice the stream of minute spiders that were running up and down along it. A glance at the wall showed a spider's 'cocoon', from which I deduced that all were the very young of a species that might well be quite big in terms of British spiders. Next I noticed that they ran in both directions and this caused me to get a lens and see just how they got around the problem of passing each other. The lens gave me the simple answer – there were two threads like twin railway tracks. It also showed me that there were six or seven spiders moving from wall to chair for every *one* moving in the opposite direction on the other thread. I examined the chair end of the threads. No spiders were coming off the end! So a steady stream of spiders was leaving the safety of their wall home and apparently just disappearing.

It took me some time to discover what was happening. At the very centre of the track, each spider in turn was crossing over the threads and then pushing them apart. On either side, other spiders acted as anchors, holding the threads close together. The central spider then released its grip on one thread and was projected into the air. It had created a bow and arrow with itself as the arrow! Having been launched, it drifted on a light breeze towards a mass of brambles and roses in a corner of the garden to start its own life in a new 'continent'. The feat was no less than those of the men who first charted the Amazon or sailed to the New World or who launched themselves towards the moon.

I find spiders *very* disturbing. They can withstand 10 times the dose of radioactive radiation that would kill a human. To accept that we will destroy ourselves in nuclear war and leave the arachnids to rule the earth appears to be a very plausible argument for those opposed to nuclear research, but it is not so. Anyway who is the spider to boast of being able to withstand this level of radiation? The bacterium *Micrococcus radiodurans* can withstand 1000 times what is a lethal dose for the spider.

A living voltmeter

Whilst on the subject of spiders, I am reminded of an article that appeared in a regular journal of a large industrial organisation which builds electrical transformers.[4] To quote less than the whole of it would be to lose the force of the argument that it brings home.

> BIG INSECT TAKE-OVER – S.A. Test Department
> It's happened. Hot off the press comes the news that the insect world is taking over, the first signs of the invasion came recently when a spider decided to take over the voltage measuring duties in our South Africa Test Department.
>
> Sammy spider was found trespassing on the Test Department's high voltage wire, so the Department's Chief Executioner was called in to apply the required 50 kV. At the crucial point, Sammy saved his skin by demonstrating his voltage measuring ability, lowering himself away from the wire on his thread at approximately 1 inch per kV. All attempts to confuse Sammy failed, as irrespective of whether the voltage was applied suddenly or gradually, he moved either with alacrity or gradually to the relative number of inches below the High Voltage wire, depending on the voltage applied. A calibrated scale placed near the wire confirmed the ability of Sammy to interpret exactly the condition on the wire. At this stage it was felt desirable to get on good terms with such an intelligent insect and a meal of flies was provided. We are standing by for the next signs of the insect take-over.

The condescending manner in which this paragraph is written is typical of the attitude of physicist and engineer alike in respect of the 'lower orders' of life. Perhaps the worst feature of all is not that the writer did not know that a spider was not an insect but that he neglected to find out which species and saw no importance in the fact that in that tiny creature was a mechanism that could replace a pair of very large metal spheres (about a metre diameter) that have been used for decades to measure a voltage by the size of gap between them in which air will ionise and break down under the stress of the voltage between them. Even though neither the writer of that extract nor I have the biological expertise to discover the mechanism within the spider, he might at least have given a biologist a better start.

The spider was in physical contact with only one 'electrode' – the high voltage wire – the other electrode being the earth including the surrounding walls of the building. How did it measure voltage so accurately? My own guess is that it could detect the stresses in the hairs on its body (spiders are often very hairy) as they were straightened by the electric stress. Making a person's hair stand on end by standing on a charged metal plate has been a standard trick in the illustration of lectures on electrostatics for over a century.

Electric fields

I was once sufficiently near to the ground end of a lightning stroke to realise that anyone standing in the open who is struck by lightning has received prior warning. I suddenly felt the hairs on the backs of my hands begin to move. They had begun to stand out at right-angles. Then the hair on the back of my neck was affected and I flung myself flat on the ground, wet and muddy as it was. A second later the lightning stroke hit the ground within yards of me.*

---

* In a lightning discharge from cloud to ground, the first thing to happen is that a current flows from ground to cloud, but the diameter of the ionised path is tiny and its duration is only a fraction of a second. This 'pilot' stroke has, so to speak, drilled a pilot hole in the atmosphere down which the main charge from the cloud can now descend, and does. The main current is of the

Some interesting work on the effects of electrostatic fields on insects was carried out in the 1950s by the West German Professor Geza Altmann. A pair of parallel plates was placed in a vertical plane on either side of a wasps' nest. When an alternating voltage was applied to the plates and raised to give a field gradient of a million volts/metre* the wasps began killing each other; they killed their queen and finally, apparently, those remaining blocked the entrance to the hive with wax and suffocated. The investigator repeated the test with other creatures including bees, ants, spiders and even mice, but failed to produce any abnormal reaction – as he put it: 'Even though the voltage was varied over a large range.'

Further experiments, which I repeated, showed that a collection of worker wasps, enclosed in a polythene box, appeared unaffected by the applied voltage, even when taken to much higher values than had been necessary to cause the 'civil war' in the nest. The German researcher went on to demonstrate that two factors, other than the applied voltage, were needed to produce the effect. They were the presence of the queen wasp and the presence of the honeycomb. His master stroke came when he placed the plates of the capacitor (for this is, in effect, what it was, with whatever lay between the plates acting as dielectric) *above* and *below* a bees' nest, as opposed to those on the *sides* of the wasp nest, and succeeded in producing the mutually aggressive effect in the bees also. What is surely significant is that having established the need for the honeycomb, the only fundamental difference between the nests of bees and wasps is that bees build their comb in vertical planes, wasps in horizontal.

Could it be that a frequency of 50 cycles/second, the commonly available commercial frequency in Europe, was in some way 'tuned' to creatures the size of bees and wasps? (Altmann varied the voltage but not the frequency.) If so, then a mouse would possibly be responsive to a lower frequency. The natural question for *H. sapiens* is clearly: 'What is *our* frequency?' The answer

---

order of tens of millions of amperes but, alas, does not carry enormous energy, at least not enough to be worth trying to collect, even though it is enough to split a tree trunk from top to bottom.

* Air breaks down electrically at about three times this gradient.

might well be in the range from 1 to 10 cycles/second. It is well known, for example, that a light flashed in the eyes of an epileptic to match the alpha-rhythm* can initiate a fit.

So far as I know, there has been no follow-up of this work, nor did it receive a great deal of publicity at the time. It could be that biological knowledge and the possession of high-voltage equipment are a rare combination. Possibly others tried and found nothing beyond what was already published. I include it here as a classic example of the gulf between physics and engineering, on the one hand, and biology on the other.

As for an insect 'take-over' of the world, it would have happened millions of years ago had they not been limited to the size they are by their breathing technique. But then, had they been bigger, their characteristic time would have put them into the same hazardous condition as those of mammal, bird and fish. That there are over a million species† and that the total weight of all the insects in the world is greater than that of all the mammals was explained concisely in an exhibition at the British Museum (Natural History) in the 1970s. I forget the precise wording but the question, 'Why so many?' was answered:

> They started early(300 000 000 years ago)
>
> They fly   (most of them)
>
> They breed fast   (an aphid averages 100 young per month)
>
> They adapt
>
> They are small
>
> They are tough   (they have their skeletons on the outside)

---

* The electro-encephalograph is a machine that can detect and measure the patterns of tiny electric currents that pass through the brain. These have been categorised roughly according to frequency. The theta waves (4 Hz to 7 Hz) are believed to be connected with enhanced mental activity, and creative thinking is believed to occur in this general range. Beta waves (14 Hz to 30 Hz) are thought to indicate logic and rationalisation. When there is little brain activity accompanied by a feeling of serenity, well-being and a totally relaxed mental attitude, the brain currents display a relatively simple pattern of between 8 Hz and 13 Hz. This is the alpha rhythm.

† Neglecting plants, worms, lowly sea creatures and micro-organisms and calling the remainder of living creatures 'animals', four animals out of every five on earth are insects.

The late J. B. S. Haldane was once asked what his experiences in life had taught him about the nature of the Almighty. He replied, 'An inordinate fondness for beetles!' (Of the one million-odd species of insect, over 400 000 are beetles.)

As usual, I digressed. We were discussing textiles. In the weaving trade we find birds, as they build nests. Crows weave twigs to make a strong platform. The birds who weave best we give the honour of being called 'weaver birds'. Beavers know the technique as they build their dams. But all of these creatures make textiles just one facet of a greater creativity. The camouflage of a chaffinch's nest, together with the warmth it conserves, can be compared to the combined skills of human builders, painters and decorators, and heating and ventilating engineers.

Basic engineering

The beavers have practised civil engineering since they became a species. It is not merely structural skill, but the interplay between a structure and moving water that they have mastered. They never needed a microprocessor nor a silicon chip. Nor will these later inventions benefit their kind – destroy them, possibly, but never benefit.

Termites are great builders – architects too. The *Guinness Book of Records* shows a termite structure, which to scale would be the equivalent of a Man-made building 4500 m (15 000 feet – nearly 3 miles) high!

From above the surface of the earth to below it, nature's engineers are legion. Moles and rabbits are the miners of the animal world. Under a lens, in the micro world there are tiny caterpillars of very small moths mining within the thickness of a leaf. They are, in fact, known as 'leaf-miners'. Some of their handiwork is a familiar sight on the leaves of garden plants. The laburnum tree, in particular, hosts a leaf-miner whose adult is a moth so magnificent (snow white, embroidered with gold), as to compete with the finest moths of the tropics, but goes unnoticed as a bit of white fluff being less than a 5 mm (0.2 inch) wing span. Imagine its impact if it were to be the size of an Atlas moth (wing span over 20 cm – 8 inches).

Engineering is seldom done by one person alone, especially when it falls into the category of 'heavy'. No one person built the Eiffel Tower. Such projects involve teamwork such as we find among the social insects. A super being might observe a team of human engineers and assume quite rightly that they were making use of what might be called 'collective thinking'. But such a description can have far-reaching consequences. In the case of humans the combined thought is the result of much conversation and much drawing of sketches and pictures. What other forms of communication might be possible?

Collective thought

It has recently been suggested that the so-called *social* insects (bees, wasps, ants), when viewed as a colony, appear to resemble a more complex animal with a single intelligence. Whether such an idea can be extended from insects to birds is an interesting question. A bird is seldom credited with much intelligence, as witness the expression 'bird-brain' often used in comedy shows as an intended insult to a human individual. Certainly I have seen upwards of 10 000 starlings in massed flight (at mating time in January/February) when the space between bird and bird was comparable to the size of one bird. I have seen such a flock turn a sharp right angle when all birds turn virtually *simultaneously*.* The same phenomenon occurs with shoals of fish. There is no leader. Even if there were, 99% of the birds could not see the leader and the message to turn would ripple through the flock like a travelling wave. One has only to see a very long column of soldiers marching three-abreast (without a band to provide the rhythm) to see a section of the column get out of step, and to see that irregular section travel from front to back, escalating in

* The correct collective noun for starlings is a 'murmuration' – a delightful name, comparable in accuracy to calling a bald man 'curly'!

length as it goes. For the soldier actually involved in the march, his only experience is that at frequent intervals he is alternately marking time or taking huge strides (about 1.5 times the statutory 30-inch military pace). If the column is very long, the 'catching up' phases involve running! It feels like the cyclist in the music-hall joke, who is said to have got his braces caught in the rear bumper of a fast-moving car!

With birds, locusts, butterflies and other migratory creatures this last phenomenon does not occur. Locusts in a swarm are so numerous that the overall appearance is that of a very dense cloud. In 1889, a swarm was estimated to contain 250 000 000 000 insects and to cover an area of 2000 square miles (5000 square km – see also p. 215).

Might it be that we have lost the ability for collective thought that may be some sort of compensation for low IQ or decentralised nervous system? Might it be that the only time we feel the effects is when we use it for evil purposes. Of the people appearing in court on a Monday morning for throwing bottles at a referee on a Saturday, a great many are very nice people. If there is one thing I fear most in this life it is the mob. We saw it on the grand scale in Germany in the 1930s. But the first time I was *really* alerted to it was when I was walking alone in the New Forest in 1948. I met a hunt moving off. There must have been all of 30 to 40 dogs and a dozen or more horses and riders. The leading dogs rushed up to me, tails wagging in obvious pleasure. I patted and stroked the leaders – did I have an alternative – then they *all* wanted to be patted. They jostled each other to get near. I was surrounded. Suddenly I said to myself, 'You had better stay vertical, boy, or the dogs on the outside just might think there is a fox in the middle!' When the first horsewoman got within shouting distance – and I *do* mean shouting – she told me to 'leave the hounds alone' and became more abusive as she approached, whip in hand. I have friends who hunt but I make no apology to them when I say that on that day I realised that fox-hunting was a barbaric sport. The Roman gladiators had more chance.

Nature is full of paradoxes. The view that a hive of bees or a nest of ants behaves like the brain of a single mammal, capable of solving engineering problems as might a specimen of *H. sapiens*, is but one useful comparison that can be made – useful because it might give us an insight into the thought process itself. On the other hand, it could be argued with equal force that such a colony, hive or nest is much nearer (biologically) to a plant, than to an animal. Remove a section of it; that which remains survives. Take 'cuttings', as from a plant, and provided you take a piece of nest containing an infant queen, both cutting and original 'plant' survive. When a nest is too big, it divides naturally, as will many plants. The analogy can be taken deep into the subject.

The following account, which strongly favours the former of the two ideas, describes an event that I actually saw for myself.

The ant engineers

The most amazing example of collective thinking I have ever seen took place one summer Sunday afternoon on the lawn of the RAF Officers' Mess at Farnborough in 1944. There was a short flight of stone steps leading down from one lawn to another. Figure 8.1 shows a cross-section of the ground and the location of an ants' nest under the top step. On top of the step lay a dead beetle that had been partially crushed by someone stepping on its rear end. It was effectively cemented to the step at that end by its own solidified juices. The ants had 'decided' they would like it for food. They had just begun to drag it towards the edge of the top step when I first noticed their efforts. About 20 ants were engaged in the pulling and there was one particular ant who did nothing but run around the beetle, stopping only for a second at a time to touch antennae with one of the pullers. The contact time of such encounters I estimated as of the order of one-third of a second. I watched one of the ants so 'instructed' and it was not contacted again for over 10 minutes. In terms of the life span of the ant, this single act of tactile communication was comparable with two humans shaking hands and in that action communicating the entire contents of this book from one person to the other.

The 20-odd ants laboured on and managed to pivot the beetle about its anchored end, but nothing more. The 'boss' ant (who

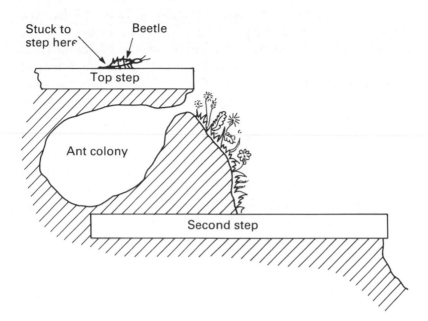

Fig. 8.1. Cross-section through an ant's nest.

was certainly not the queen and, so far as I could tell, no different from any other worker ant) began to move faster, then touched another worker which sent the latter scurrying back to the nest. A few seconds later another couple of dozen ants emerged and grabbed the beetle. All the ants heaved in the same direction but the beetle was stuck firm. The boss ant, in a near frenzy, ran around and all over the beetle, then, in the space of a few seconds, contacted several of the workers who ceased pulling and proceeded to gnaw at the beetle's rear end with their mandibles until they had severed it from the step. (Unlike what humans might have done in a comparable situation, the other 30-odd did not fall on their backs at the sudden release!) Within half a minute the beetle was poised on the edge of the step.

'Are *they* in for a surprise,' I thought. 'As soon as the centre of gravity is over the edge, down it crashes on to step number two and they will never lift it back through all those weeds growing out below the nest entrance.'

The boss ant 'detailed' another few ants to return to the nest. A hundred or more ants emerged and all seized the beetle at the end remote from the step and pushed the beetle right to the point of balance on the edge. I was suddenly aware of a great mass of ants, all holding on to each other, streaming out from the nest, and those in front made their way up the edge of the step and grabbed the beetle. In a flash, all the ants on top of the step released their grip. The beetle fell, as shown in Fig. 8.2, not down to the next step, but only an inch or so below the nest, and in a matter of seconds the horde had heaved the beetle indoors. It had travelled the minimum distance in the process.

What a feat of engineering! What a communication system; what economy of ant power – never more ants than were needed for any phase of the operation. The principles of leverage, centre of mass, friction and inertia were obviously being 'appreciated' by creatures with no brain, poor eyesight and a controlling chief engineer who, on the face of it, was no different from any of the other workers – all neutered females. This situation was unique for them and for their generation. Could this ability really have been handed on from 20 generations ago when last a similar

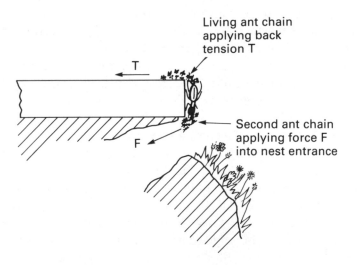

Fig. 8.2. Solving the crushed beetle problem

situation occurred? Where did they learn their engineering, their ingenuity, their project assessment? What was special about the boss ant and how was she chosen? Just what *can* be passed on in the genes of so relatively simple and primitive a creature?

There were a score of questions and no answers. Maybe the man who wrote the book of Proverbs had had an experience similar to my own.*

Engineering techniques

It is as easy to find, in living things, the individual processes that make up the whole of engineering, as it is to find the process as a whole.

The basic *cutting* processes used in lathes, milling machines and drills involve ripping off thin layers of material with a hard, sharp edge, which theoretically only comes in contact with the work piece at the start of the cut. This is precisely the way in which wasps strip wood to make paper. Paper making, incidentally, was a trade long before humans appeared on earth.

Shearing as with scissors, is another well-known cutting method which also exploits a common weakness in materials to fail in shear when their tensile strength is very much higher. Shearing is a common process in mammal, bird and insect alike: mandibles, teeth and beaks all do it. These implements are often both shearers and crushers in combination. But the locust's mandibles can eat through metal! Wasps will catch bluebottles in flight and bite off their heads, and the fly's neck is almost wire-like.

Drilling holes in wood is easy for insects but their methods are not those of the twist drill. At least one creature, however, is aware of the technique of drilling a small 'pilot' hole before attempting to drill one of large diameter. The large drill will follow the pilot hole and give greater accuracy than it would if it had to maintain its own centre. Inspection of the way in which a

---

* 'Go to the ant, thou sluggard; consider her ways, and be wise.' (Proverbs, vi, 6.)

twist drill is sharpened reveals the reason for this. The creature that uses the pilot hole technique is yet again the fearsome arachnid. A tarantula will pull a hair out from the skin of its victim in order to make a pilot hole which it can then enlarge. Ingenious? No, just evolution. But there must have been a *first* one, just as there must have been a first solitary wasp to pick up a tiny pebble in its jaws and use it as a ram to flatten the walls of its earth burrow (mostly they still use their heads). (See also p. 218.)

The British goat moth larva is a wood-boring insect. It is so-called because it smells like a goat! The smell is intentional, a built-in atomiser (another human 'invention' a few million years late) to spray foul acid to dissuade birds from eating it. I know a man whose timbered house had goat moth larvae in the beams of a ceiling. Unlike most moths that operate on a one or two generations a year basis, the goat moth feeds up as a caterpillar for 2 years. The man decided to wall them in by putting a sheet of zinc around the beam. When the time came for the moths to emerge, they produced among them (presumably on some sort of collective motivation) enough acid to dissolve the zinc. One worries again about the arachnids and their radiation-resistant ability, as yet unused (see p. 226).

Adaptation

This subject, so important to the continuation of species, having been omitted from earlier chapters as a separate subject, must follow the above story, which itself is an amazing example of the phenomenon. So is an experience I had as a boy when I spied a fully fed puss moth caterpillar walking across the pavement. Not knowing what it was, I took it home to put into one of my moth breeding cages. I knew not how big it would grow, what it needed for food – nothing. My father, looking ruefully at his cabbage patch said that most things seemed to eat cabbage, so cabbage was offered in the cage and I went off to school. On the way home I met a local naturalist who showed interest in seeing my caterpillar. By the time we reached home it had pupated completely. Now, the puss moth spins a cocoon on the bark of a tree and attaches bits of bark to the sticky outside so that the whole thing becomes totally camouflaged (so the moth is clearly a house painter and decorator, by trade). Having nothing but cabbage leaves to work with, my larva had chewed up the cabbage, mixed it with saliva and made a paste with which to house itself, much as wasps do with wood. 'Poor thing,' said the expert, 'the cabbage will putrify. You have lost your moth.' Three months later a perfect moth emerged quite normally. The cabbage had not putrified. The saliva had seen to that. This was no accident. It was just another example of the amazing powers of adaptation.

In the 1950s a lot of work was done in studying the peppered moth. In its pre-Industrial Revolution days it was off-white peppered with black. But in places like industrial Lancashire, more and more specimens emerged totally black. Biologists argued as to whether this was natural selection taking place as the result of the pale specimens easily being picked out when at rest on tree trunks by birds, or whether they became black through eating so much soot on all the leaves of their food plants (the rose family). However, came the smokeless zones (the Clean Air Act of 1956), and the question was: will the original form of the peppered moth return to Lancashire? It did! Dramatic evolution within a human lifetime, for whatever reason.

The glorious 'birdwing' butterflies of New Guinea feed on the juices of rotting animal flesh and animal excreta. (It appears that the more beautiful the insect, the filthier are its habits. A species with the biological name *Euthalia dirtea dirteana* is gorgeous, and to show that classifiers have a sense of humour a related species is named *E. dirtea khasiana*!) David Attenborough, in his magnificent TV series *Life on Earth*, showed flowers that smelt like rotting flesh and whose surfaces were wrinkled like dead skin. Adaptation is by no means confined to the animal kingdom.

But adaptation cannot always come to the rescue. Although rabbits have adapted to the disease syphilis, a single European man went exploring in the Amazon Basin, found a tribe new to geographers and anthropologists, developed a good old English common cold and wiped out the tribe in a few weeks.

An example where adaptation seems to be taking a long time

to happen is the way in which blackbirds dip low in flight as they cross roads. The reason for dipping is to avoid attack by hawks, for it would be a foolish hawk that dived from 100 feet (30 m) on to a bird flying only a foot from the ground. It just might miss! But we have had close on a hundred generations of blackbirds who have known cars and there are far more cars than hawks and yet blackbirds are slaughtered by the hundred thousand annually on the roads of Britain.

One of the best examples of a human ability to adapt consists of an experiment in which volunteers wore spectacles which gave them an upside-down view of everything. They had to learn how to live normal lives under the handicap. After some weeks, each volunteer in turn would suddenly see things the right way up again. Not gradually, but suddenly, at first for a few seconds only, then for longer and longer periods until adaptation was complete. When they then removed the spectacles they saw the world upside-down with their naked eyes, until they had re-adapted.

One could obviously write a whole book on adaptation, but I will end this section with just two stories. A sporangiophore (a type of fungus) grows at an enormous rate (inches per night) and will direct its growth towards any visible light source. It can detect and react to a single photon. More than that, it can adapt its sensitivity to minute changes in illumination intensity against changes in background brightness over a range of 1 000 000 000 : 1.

In his book *African Genesis*, Robert Ardrey relates a personal experience that must have a claim to being the ultimate piece of adaptation.[5]

> There is a creature native to Kenya called the flattid bug, and I was introduced to it in Nairobi, some years ago, by the great Dr L. S. B. Leakey ... But to speak more precisely, what Dr Leakey introduced me to was a coral-coloured flower of a raceme sort, made up of many small blossoms like the aloe or hyacinth. Each blossom was of oblong shape, perhaps a centimetre long, which on close inspection turned out to be the wing of an insect. The colony clinging to a dead twig comprised the whole of a flower so real in its

seeming that one could only expect from it the scent of spring.

My real moment of astonishment, however, was yet to come. I had never seen anything comparable to the insect-flower before, but such protective imitations exist widely in nature . . . How random mutation can account for such imitations must be left for geneticists to worry about. But imitation exists in the natural world, and to demonstrate my sophistication I expressed my admiration for the flattid bug, but threw in a few comparable examples.

Leakey listened with amusement, and agreed with me, but then mentioned an off-hand fact. The coral flower that the flattid bug imitates does not exist in nature. And my moment of speechlessness began. The flattid-bug society had *created* the form.

While I was suffering mental indigestion from the extraordinary statement, the eminent Kenyan . . . now contributed further material to my flattid-bug bewilderment. He told me that at his Coryndon Museum they had bred generations of the little creatures. And from each batch of eggs that the female lays there will always be at least one producing a creature with green wings, not coral, and several with wings of in-between shades.

I looked closely. At the tip of the insect-flower was a single green bud. Behind it were half a dozen partially matured blossoms showing only strains of coral. Behind these on the twig crouched the full strength of flattid-bug society, all with wings of purest coral to complete the colony's creation and deceive the eyes of the hungriest of birds.

There are moments when one's only response to evolutionary achievement can be a prickling sensation in the scalp. But still my speechlessness had not reached its most vacant, brain-numbed moment. Leakey shook the stick. The startled colony rose from its twig and filled the air with fluttering flattid-bugs. They seemed no different in flight from any other swarm of moths that one encounters in the African bush. Then they returned to their twig. They alighted in no particular order and for an instant the twig was alive with the little creatures climbing over each other's shoulders in what seemed to be random movement. But the movement was not random. Shortly the twig was still and one beheld again the flower. The green leader had resumed his bud-like position with his varicoloured companions just behind. The full-blown rank-and-file had resumed its accustomed places. A lovely coral flower that does not exist in nature had been created before my eyes.

More engineering

Without elaborating, the following is a list of techniques common to modern engineering and to living creatures.

The ball and socket joint is to be found at the base of every spine of a sea urchin. It is also to be seen in the joints between human bones. The zip fastener was extended to use stiff bristles which interlocked two surfaces, one such successful product being marketed under the name Velcro. This same mechanism is used to link the fore and hindwings of certain species of moth.

The screw motion of a ship's propeller is to be seen among certain pupae (dead though they appear for most of the time) as they burrow into soft soil or leaf mould in which to spend the winter. Many species of bacteria move by a screw motion of their tails – 'flagellation' is the biological term. The process has been well researched.[6]

Hawk moths, so called because of their ability to hover like hawks, send their long probosces into flowers such as nicotiana. 'Refuelling in flight' was what we termed it when it was first performed between fighter and parent aircraft. Jet propulsion, the great British invention of World War II days, is practised by the larvae of dragonflies who live under water in ponds and lakes. Equipped with powerful mandibles and legs but with nothing to equal small fish in swimming ability, they attain enormous accelerations by drawing-in water and ejecting it backwards, as from a hosepipe. A few inches of such accelerated motion is all they need to grab a small stickleback. The same action is used by the squid, as highlighted by David Attenborough.

Gyro stabilisers in ships, aircraft and spacecraft involve the use of spinning wheels, but an oscillating mass has all the properties of a wheel and mathematically may be so represented. The common crane-fly ('Daddy long legs') is strictly peaking a four-winged insect, but the pair of hindwings has degenerated (?) into a pair of stalks carrying a knob on the end of each. These vibrate and act as gyro stabilisers.

Travelling wave motion, as used in induction motors and electrical machines generally, can be made visual by a mechanical model, such as that shown in Fig. 3.1. This type of motion is to be seen in the legs of millipedes, centipedes and the like and in more sophisticated, continuous sheet form in cilia and in many aquatic animals from molluscs to flat fish to dolphins.

The 'micro-switch' and like gadgets depend on putting a piece of metal under stress until it 'flips' into an entirely new position. The idea is to produce a sudden, highly accelerated motion as the result of a microscopic amount of initiating motion. Plants manage the action without resorting to spring steel. Its main use is for the dispersal of seeds by hurling them through quite large distances. The 'clicks' coming from the seed pods of lupins, crane's bill, castor oil plant, broom and many other podded plants mark the attainment of the 'top-dead-centre' position necessary for the action.

The action of a siphon can be quite surprising when you first encounter it. A simple version is shown in Fig. 8.3. The beaker empties itself as if there was a pump inside the tube. All that is necessary is that the exit end of the tube should be below the water surface in the beaker, and therefore below its base, if it is to empty it completely. It is only an example of the law of minimum energy, and yet quite a striking effect. It is used by shellfish of many species, one of the best known being the pearly nautilus (*Argonauta argo*).

When it comes to structural engineering techniques, botany dominates. Tubular constructions are to be seen everywhere, giving maximum strength per weight against bending moments. But among the incredibly diverse insect forms, at least one (a parasitic caterpillar) also uses the techniques.[7] Corrugated material is also extremely common in plant and insect life; stems, leaves, wing

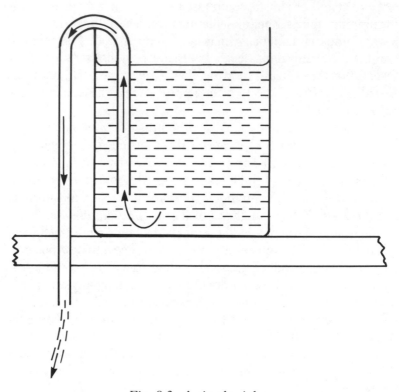

Fig. 8.3. A simple siphon

cases of beetles all need the extra strength it provides. Reinforcement by the use of fibres is illustrated in Fig. 8.4, which is taken
from Felix Paturi's book, *Nature, Mother of Invention*,[8] in which
greater detail is given to botanical aspects than space will permit
here. Among the fascinating comparisons made is that the original
design of the structure of the Crystal Palace, destroyed by fire in
1936, was inspired by the leaf of the huge South American water
lily, *Victoria amazonica*. For the structures of cranes and their
operating cables one should study the giant spider crab from
Japanese waters. Its legs reach the incredible span of 5.8 m
(19 feet). It would perhaps not be 'leaning over' to regard the
archer fish, which ejects jets of liquid at flies sitting on rocks to
bring them down into the water, as having a built-in water

Fig. 8.4. Reinforcement by fibres – the skeleton of a dead seguaro cactus has exactly the same arrangement as the cage of a concrete column.

cannon? Ants can similarly eject jets of formic acid when threatened.

Among nature's engineering mysteries that we have not solved is how a seagull desalinates sea water by means of a gland, and how a hawk positions itself when it hovers. Taking the seagull first, imagine how many lives might have been saved at sea, if only we knew how to make a device as small as that of the salt gland that would deliver fresh water at the same rate. With regard to the hawk, we need only look out of the window of an aircraft when banking and try to decide on which point on the ground is directly below us, to see how difficult it would be to hold a position relative to the ground in a 30 mile an hour wind. The hawks do it, so what do they use as their measuring device?

In acoustic engineering it is not surprising to find the insects way ahead. The male mole cricket, *Dryllotalpa vinaea*, researched by Bennet-Clark,[9] digs a double exponential horn, as shown in Fig. 8.5. Lower down, beyond where the tubes join, the tunnel is enlarged to form a bulb. The twin horns act as an acoustic transformer and concentrate the love call of the male into a vertical, semi-circular disc-shaped beam to attract as many passing females as possible. The walls of the burrows are extremely smooth. The bulb acts as an acoustic tuning device to give the whole system a high Q-value. The result of this is that the sound emitted can be heard by the human ear at a distance of 600

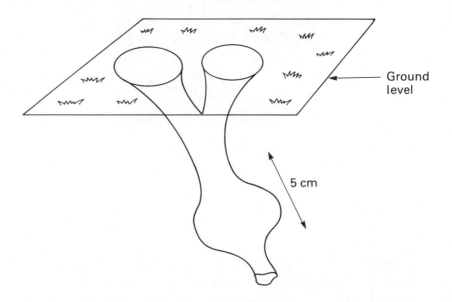

Fig. 8.5. Cross-section through the burrow of a mole cricket.

metres (2000 feet)! The note is almost pure (between 3400 and 3500 Hz, which is 3½ octaves above middle C on a piano). Bennet-Clark observed, in relation to the purity: 'If moonlight could be heard, it would sound like that.' The efficiency of the cricket's acoustic system is about 35%, far greater than that of commercial loudspeakers. Its output is only 0.0012 watt. It is the shape, and shape alone, of the tunnel system that gives it the carrying ability.

A South African tree cricket, *Oecanthus burmeisteri*, turns itself into part of a microphone. It gnaws a pear-shaped hole in a suitably sized leaf and sits in the centre of it, holding the edges of the hole with some of its feet and pressing against the periphery of the hole with its wing cases. The frequency here is about 2000 Hz. The work was reported in the *New Scientist* in 1975 and published in full in *Nature*.[10]

These acoustic examples might well have been included in Chapter 6 for they are both essentially concerned with shape. Most fundamental of all shapes in human progress, the plough blade (see p. 30) is easily identified with the shapes of leaves. Another technique, well used by plant, animal and engineer is that of thermal insulation. Warm-blooded mammals (seals, sea lions, polar bears and whales) that live in ice-cold waters certainly need it. Their methods are basically the same as those of Eskimos (who after all, use the animals' fur as one method).

Still more trades and professions

The famous angler fish dangles bait (without hook) within jaw-snapping distance for prey foolish enough to be tempted. As for fishing as we practise it, the idea of a hook barb emerges in nature not in that context but in the sting of a bee.

Ants provide us with still further human-like activities. They sow seeds with a view to reaping the grain, they reap crops of fungi, milk 'cows' (aphids that will exude a 'desirable' fluid when stroked – desirable perhaps in the way that whisky is for us) and tend 'animals' (large blue butterfly larvae).

Where mimicry is involved, moths imitate wasps, beetles, spiders, butterflies, owls (and even small mammals). The acting profession is well represented. So also is the nursing profession among the social insects. Swallows and house martins are expert plasterers. They carry mud from river banks and build hard outer shells to their nests under the eaves of houses. One of the oldest professions, referred to from Biblical times, was that of tent maker (Saul). The larvae of the comma butterfly (and others) are certainly just that. They communally erect a tent of silk over their entire feeding area, partly as camouflage defence against birds and partly as shelter, presumably.

Migrating birds, insects and fishes are the world's best navigators. The feat of monarch butterflies in America has already been discussed (p. 220). It has been suggested that British-based eels find their way to South America and back by homing in on the temperature of the Gulf Stream. Moths that find their mates over distances of miles have been shown to be micro-wave radio telegraphists.[11]

Yet among the entire natural livestock of the world there appears to be no evidence for the existence of accountants, lawyers, politicians or Inland Revenue inspectors – there must be a moral there somewhere!

Among other classes of human, not divided by profession but by activity, we can find murderers (it is rare that wild creatures kill their own kind, but queen wasps will murder young queens in their cells), cannibals (some caterpillars will eat a leaf and continue through the caterpillars they encounter on the way, but I regard this as mere carelessness!) and slaves. All worker bees, wasps and ants could be so described, having once been potential queens.

There is just one further point about murderers before going on to examples of pure science. One of the most significant experiments of this century, I call the 'rats per square yard' experiment. A number of rats were kept inside a limited area compound, well fed and housed. The number of rats was gradually increased week by week and beyond a certain 'rat density', they turned on each other and fought to the death. How far along this road is *H. sapiens*, and what has the phenomenon in common with football riots, racial riots and more recently riots for their own sake?

Purer science

Having disposed of engineering techniques, the more 'exotic' sciences have even more spectacular counterparts in living tissue, or even in minerals.

Fibre-optics, for example, is an exciting new industry on its own. But the mineral xylenite is fibrous in structure and conveys light from one polished surface to another, however thick. Whilst it could be argued that this is 'nature's' fibre-optic, it was never *alive*, so evolution cannot be invoked, nor is any benefit to be assigned to it. Let's face it, it *is* an accident, and this single example suffices to throw doubt on the logic of evolution of this or that mechanism, *however obvious*.

The plant fibre-optic occurs in what Walt Disney called 'the living desert', in Arizona. Cacti have remarkable survival devices, water storage, spines and so on. But they get eaten anyway if they are the least bit green. So some evolved to look like pebbles on the sand. They are remarkable, but presumably get eaten too, for other species went completely underground (12 cm (5 inches) under the sand). Now, being plants, they need sunlight to enable photosynthesis to take place. So they send up a multitude of thin fibres from the surface. Since very little light is transmitted through a 12-cm (5-inch) block of ordinary glass, these fibres have at least a part of the properties of fibre-optic strands.

The electric eel's use of radar is well known, as is the acoustic radar or 'sonar' of the bat. But less well known is that within a few years of the invention of the laser, that made holography practicable, two electrical engineers, Gamow and Harris, writing in the Institution of Electrical and Electronic Engineers' journal *Spectrum*,[12] on the subject of this book – 'What engineers can learn from nature' – mentioned almost casually as an example of their main theme:

In most biological systems, adapt or perish means just that.
Thus, the bat had to learn to survive in the environment of
a dark cave, and under this pressure not only evolved a
mechanism of echo location but also what appears to be a
system of acoustical holography.[13]

As regards the sonar part of this creature's make-up, an Italian
named Spellanzani did experiments with bats in 1799, and found
that when he blocked their throats they fell to the ground. At
that time it was reckoned to be understood that bats avoided
obstacles during night flying by possessing amazing tactile recep-
tors on their wing tips that 'felt' the approach of a solid object
through the viscosity of air (and had a response time of about
0.001 second, apparently!) Poor Spellanzani, whose theory was
that the bats issued shrill cries which were reflected as echoes,
was totally disbelieved. More than that, he was publicly ridiculed
by the leading biologist of the day who exclaimed, 'If then, they
see with their ears, what do they hear with their eyes?' Today we
might give him the answer: 'Well, they see, but you would never
believe their *other* purpose.'

It took 120 years before echo depth sounding became well
known and a classic paper on the bat's sonar was re-discovered.

Now let us talk thermometers. Gamow and Harris pointed out
that in many snakes their infra-red detector 'is able to distinguish
unerringly 0.001 °C from ambient' and that they use this ability
in two ways (as usual!). They can locate and capture warm-
blooded prey by pure heat radiation, and they can detect micro-
habitats of the correct temperature to suit their own bodies, so
they know exactly where to spend most of their time. They added
that our heat sensors may be potentially as sensitive as that of a
boa constrictor but 'they are buried under 300 μm of flesh'
(0.3 mm or 0.012 inch).*

It has also been estimated that the common bed bug who finds
his host by infra-red radiation can detect a temperature difference
of 0.003 °C between one end and the other of his own proboscis.

It has taken us a long time to produce 'cold light', as produced

---

* 'I shall return to Gamow and Harris' message later on p. 257.

by glow-worms and fireflies, but at least the mechanism is understood. Many insects produce endothermic reactions and every Victorian bug collector knew that to detect whether a pupa was alive or dead you placed it against your lip (your most accurate built-in thermometer). If it felt 'ordinary', it was dead. If it felt cold (like metal in winter) it was alive.

Trilobites' eyes correct for spherical aberration, an optical technique only a few centuries old in our history. Water beetles, bees and other insects have a built-in polarimeter which enables them to detect reflected light from the surface of water on the one hand, or detect the plane of polarisation of light from the sky and hence deduce the position of the sun. Thus they can locate the honey fields by the famous 'bee-dance' technique researched so well by von Frisch a few decades ago, even though meteorological conditions are 10/10 cloud. (See also Chapter 6, p. 146.)

There are a few butterflies that carry a diffraction grating on their wings. The sub-family *Pierella* contains a number of species such as *hyceta*, *rhea*, etc., all of which are brown butterflies about the size of the English 'cabbage white'. Photographs from different angles all within a 3° arc reveal all the colours of the rainbow. These colours are undoubtedly produced as a first-order diffraction spectrum. The question is, why? I doubt whether colour has anything to do with the mechanism at all. A thousand people will pass set specimens of these butterflies in a museum without a single person seeing the colours at all. In flight they would be virtually invisible, occupying the correct angle with respect to the observer for little over a millisecond at a time. What is certain is that the wing structure that caused the colours is a set of very uniformly divided, straight demarcation lines. It is highly likely that these scale arrangements have an entirely different primary purpose and that colour production is incidental. It has been suggested to me that there are parallel lines on each scale of the butterfly. This is true of most of the 160 000 species of Lepidoptera, so it cannot be responsible for the phenomenon associated only with *Pierella hyceta* and its relatives. The same can be said of the gorgeous blue South American butterflies of the Morpho family. (Incidentally morpho means 'shapely', not necessarily gorgeous.)

Numerous biological texts jump to the conclusion that the

colour arises from the scale *structure* as opposed to a pigment, for whereas 'ordinary' butterflies kept in a frame on a sunlit room wall will be bleached white inside 30 years, the blue Morphos stay just as bright for 1000 years. This deduction is correct. But most authors go on to state with authority that this is an optical *interference* effect, pointing out that each tiny scale is hollow and the top surface is semi-transparent. Light reflected from the top surface, they explain, interferes with light from the bottom surface to produce the colour. If this were true, it means that the space in every scale (and there are about a quarter of a million scales per butterfly) must be identical in thickness to an accuracy of better than 10% of the wavelength of blue light.* In nature we would expect to see all the colours of the rainbow, as when we see colours in oil floating on water. But it is not so. There are many species of blue Morpho and you can identify the species by the shade of blue. In any case, you may have a size range of 2 : 1 within a species, depending on how well they fed as caterpillars.

The colours of the Morpho butterflies, the day-flying Madagascan moth (*Chrysiridia madagascarensis*), the spectacular colours of many tropical (and a few British) beetles, the 'luminescent' colours in the feathers of humming birds and other feathered species, and of tropical fish such as neon tetra are all examples of nature's 'liquid crystals'.

Now, liquid crystals are a very 'second-half twentieth century' device. We use them as display units in pocket calculators. They change colour with applied electric stress, or with temperature. But does it seem likely that the blue Morphos have ever changed colour? If so, no-one has ever seen it happen. Could it be that we have still to discover the more useful properties of liquid crystals?

From optics to magnetism, termites appear to have a built-in compass. There is an Australian species, *Holotermes meridionalis*, that in fact is known as the 'compass termite'. It lives in huge 'colonies of colonies', where each compound reaches as much as 3.3 m (11 feet) in height and there may be tens of these com-

---

* 4600 ångström units, i.e. 0.00046 mm.

pounds in one site, resembling a kind of ghost city. Each tower block is shaped like a wedge of cheese, but the wide faces are always located to face East and West exactly. Rather than magnetism, however, it seems more likely that their aim is to keep as much of the outer wall in sunshine for as long as possible. But how did they decide on the minimum apex angle that would not collapse, other than by inherited characteristics (Lamarckism) and collective thinking?

Magnetic domains in a metal have a remarkably similar form to that of 'brain coral'. Whether this is an accident or can be traced to the basic building unit of each – a tiny living organism in the case of coral, a tiny crystal with dipole properties in the case of metal – has never been confirmed, so far as I know.

Techniques in medicine and surgery

If I knew more about these topics, I could doubtless find dozens of examples. The few I know might suffice.

Human blood normally contains a coagulant to stem the flow from a wound. People with too little are said to suffer from haemophilia. People with too much develop heart trouble due to clots forming within the circulatory system. The remedy for the latter is the injection of the right amount of anti-coagulant. This same property of the blood suggests that it will impede the mosquito when it tries to suck human blood through the tiniest of drinking straws. It would, but for the fact that the mosquito first injects its own anti-coagulant, using of course a 'hypodermic needle' (I had just called it a drinking straw). If you swat the beast at this stage – and most of us do – you may develop a red lump, if you react to the anti-coagulant. But if you can resist the urge to swat and have the time and patience to let *Culex pipiens* complete his feed, he will suck out all the anti-coagulant he put in, plus a tiny drop of your blood, and you will develop no lump – a small price to pay, if you have time! Other examples of hypodermics, of course, are nettle and wasp stings, but their needles are more flexible and painless than those of our GPs (if only the chemicals injected were not painful).* Spiders and some of their predators use anaesthetics frequently. They insist on having *fresh* meat, as do most carnivores including ourselves. Some of us say we should not kill animals at all. Others will quote the New Testament and God's own authority – 'Rise, Peter; kill, and eat.' However we live, we have a strange inherited loathing for creatures that eat flesh from animals that have died from natural causes. Yet these creatures clean up the earth; surely they give us incredibly good 'room service'? See how we use their names to hang them on to people we hate: 'jackal', 'vulture', 'crab' and 'crow' are hardly complimentary names.

The spider-hunting wasp goes in search of a spider bigger than itself. It must inject the spider with anaesthetic on its underside. Now how do you get at the underside of a huge spider without retaliation? The wasp must walk the last part of the assault with its abdomen bent under its thorax and emerging from between its front legs, and one cannot imagine a more vulnerable position. But it wins nearly every time. A magnificent film sequence of the fight was captured for Disney's 'Living Desert'. The anaesthetised spider is then dragged off to provide fresh meat for the wasp larvae when they hatch.

As a boy, I cut myself, was bruised playing rugby, suffered the usual collection of pulled muscles, cramp and the pains associated with pleasure, i.e. games. I wondered how a more severe injury, like an amputated finger or hand, could be withstood, if the pain increased in proportion. At the age of 26 I had a motor cycle accident and removed a piece from my leg bone (tibia). Then I knew. In the case of serious injury the tissue provides its own 'local anaesthetic' and at the time of the injury you feel nothing. (The real pain comes later!)

In the 1970s a brilliant surgeon named Patrick Steptoe successfully 'planted' fertilised eggs in the human womb. This practice has been common among the parasitic flies, many of which

---

* The common housefly (*Musca domestica*) can appear to 'bite' humans, notably on the legs and through ladies' nylons moreover. What they really do (like the fierce dog in the after-dinner story who had lost all his teeth) is to give you a nasty suck.

selfishly plant their own fertile eggs inside caterpillars and the emerging larvae eat the caterpillar alive, slowly. 'Nature in the raw is seldom mild' said a tobacco advertisement in the days when smoking was regarded as a healthy thing to do and 'strong' tobacco was a status symbol for manhood.

Weapons

No comparison between *H. sapiens* and anything else would be complete without a discussion of violence and destruction and the means thereof. There are a few species besides ourselves that habitually kill for the sake of killing (not their own kind, of course). They include tigers, magpies, possibly wasps and doubtless others. But in the main, nature's weapons serve one or both of two purposes:

1.   to provide food;
2.   to aid survival.

We can begin with straightforward physical blows. The horse and his relatives, such as the zebra, deliver them with their hind legs. This method has disadvantages in that you cannot face and therefore observe all the actions of an enemy, but at least it offers a truly Partheon shot to those pursuing. Kangaroos and rabbits use their hindlegs also, whilst among large birds the swan will attack an intruder with its wings as fists.

Biting is the commonest form of inflicting injury in defence or of killing for food. It is used by mammal, bird, reptile, fish and insect alike. It is very effective. A beetle about to be swallowed by a frog will bite the tongue and cause the frog to eject it rapidly. The bite is the sling of David against the Goliaths, as well as the ultimate natural weapon, when used by *Tyrannosaurus rex* and all who followed him.

Poison is a common defence mechanism, generally built into the body of the intended victim in the cases of insects and of plants. Injected poison can be used either for defence or for attack. Used by snakes, jellyfish, shellfish and plants, it can be intended to be lethal or merely anaesthetic, as we have seen. In milder forms it becomes simply a bad taste or smell, as in skunks, beetles, etc. But what is seen as 'bad' varies from species to species. Who but carrion birds and butterflies would regard rotting flesh as an *appetising* smell?

Primitive humans made sharp-pointed weapons which have survived in part even to the present day. The horns of mammals perhaps showed us the way in the Stone Age. There are spiny fish, including the one that inflates itself in the mouth or gullet of a predator – a most ingenious and clearly most painful weapon – spiny mammals (hedgehog, porcupine), spiny seashells and spiny vegetables by the thousand. But there are also other creatures for whom the spines hold no terror. Baboons will eat prickly pear and cacti generally. Horses can withstand fairly stiff thorns.

Among the less common weapons are the ability to strangle, in which the boa constrictor and anaconda excel, to electrocute (and there are several species of fish besides the eel that can do this) and to hit by water jet (see p. 244). One of the simplest and least-expected mechanisms was also illustrated in the Disney film of the desert. Small rodents were being threatened by a sidewinder snake. They appeared unconcerned, even unaware of the approaching danger. But when almost within striking distance, they turned their backs on the snake and kicked sand in its eyes. The snake retired, hurt. It had no eyelids!

Acting can also be included as a defence mechanism. Feigning death deters most creatures bent on feeding and is practised by a wide variety of species. This is hardly a weapon, and can be classified with the ability to run fast, to fly fast, to mimic a more fearsome creature and to hide by camouflage. Among the mimics might we include the great apes who used humans as their models?

Mimicry in reverse

I return to the article by Gamow and Harris referred to on p. 250. They ended the paragraph on the bat with the words: 'Indeed the efficiency with which a bat uses acoustical holography to recognise the shape of objects – that is, to identify prey – is still far superior to that of devices built by man; thus, we should still be learning from the bat.' To this I would only add: and the beetle and the bedbug, the elephant, etc. . . . etc. They continued. 'Similarly, if one could discover the nature of the glue with which barnacles bind themselves so avidly to surfaces, even inert ones such as Teflon, and bottle this substance, one might reap a fortune.' Someone wrote a sentence purporting to have been said by an elephant: 'We are the keepers of secrets for we walked the earth when it was young.' This would be more true of trilobites and their surviving relatives, surely?

The messages of Gamow and Harris are the messages of this book also:

> Why have living systems succeeded in so many areas whereas man's technology, to date, has failed? The reason for this success lies in a situation of pressure for evolutionary survival, a situation similar to that of today's professors who must publish or perish and engineers who are required to design better and better mouse traps.

They then paraphrased two quotations from Max Delbrück:*

> Trying to discover how a biological mechanism works has an advantage over solving problems in nonbiological areas since one is sure the problem can be solved; we must just be clever enough. A physicist, for instance, may spend a lifetime trying to construct a superconductor at room temperature without ever discovering that it is impossible. Nature's puzzles, on the other hand, have been *done*; they just have to be unravelled.

It is often believed that Buckminster Fuller got his ideas for the geodesic dome from the studies of viruses by Caspar and King. In fact, it happened the other way around, and those authors themselves stated: 'The solution we found was, in fact, inspired by the geometrical principles applied by Buckminster Fuller in the construction of geodesic domes.' Actually, this is

---

* Nobel Prize Winner for Medicine and Physiology.

the *usual* way around: discover sonar and *then* find out about bats; discover radar and *then* look at electric eels; discover fibre-optics, holography, liquid crystals . . . The message in this paragraph is simply that it is time we reversed the process. We were not even good enough to see the principle of the underslung chassis in beetles, spiders etc. We had to have a generation of overturned horse carriages to make us see it.

In the eyes of insects, because they are so small, there are interference filters, diffraction gratings, multimode waveguides and variable attenuators. John Lenihan put the message concisely in 1972 when he wrote of the human ear:[14]

> To Helmholtz the ear was a harp – a system of resonators
> allowing the analysis of complex sounds. Later it became a
> telephone (with the auditory nerve as a cable and the   .
> frequency analyser hidden in the brain) and more recently
> was found to embody such technical novelties as impedance
> matching, frequency analysis and automatic gain control.
> The brain, once thought of as a telephone exchange and then
> as an electronic computer, now turns out to be a holographic
> data storage system.

What will be next? Apparently we have to wait for the pure scientists to tell us. When will we learn? It was clear to Leonardo da Vinci as he said, 'Fools, open your eyes. It is all there to see.' The Book of Proverbs tells us: 'Wisdom is the principal thing; therefore get wisdom.' In the 1990s, for 'wisdom' substitute 'biology'. Lenihan concluded in 1972, after stating how the reflecting layer at the back of the retina allows nocturnal creatures to see in the dark, that it

> occurs in very sophisticated form in the sea trout found in
> the Gulf of Mexico. In these fishes, the reflecting layer is
> closely packed with small spheres of fatty material which act
> in much the same way as the microscopic beads in
> 'Scotchlite' reflecting tape.
>
> Observations such as this would have been applauded as
> evidence of the divine omniscience. Today the impact is less
> profound. The body of even the simplest living creature is a

structure of great complexity which we shall never fully understand but which we may describe in a variety of ways, according to the state of knowledge at any particular time. Physiology, like history, must be re-written in every generation if it is to be more than a collection of anecdotes.

Let us then proceed . . .

9

Into the complex

The year 1951 saw the beginning of a new technological era that was to influence all our future generations. The first full-scale, commercially built, digital computer was launched in Manchester. Almost at once the question: 'Can it think?' occupied philosopher and journalist alike. The centre of the argument was the ability of the computer to generate 'random numbers' (see also p. 219). At first sight it is easy to produce random numbers *without* a computer, surely? You only have to put discs numbered 1 or 2 or 3 etc. into a hat, shake them up, and draw them out blindfold, as is usually done with raffle tickets at a garden fete. Apparently not! Well then, just write down 10 digit numbers as they come into your head. Here is an example: 3619950427. This is quite genuine. I wrote it down just as it came to me in the flow of writing. On examination I found it fairly typical of the kind of non-random number that the pre-influenced human brain will produce, in that:

1. There is a predominance of odd digits over even ones: the word 'random' suggests 'disorder'; *even* numbers suggest symmetry and therefore 'order' (in the example I wrote 6 odd numbers to 3 even).

2. There will be few zeros (the only reason that *I* included a zero was that I was aware of the probability of not doing so).

3. There will be few pairs of the same digit next to each other, still fewer groups of three digits of the same kind, and so on (again, I wrote two nines because I was aware of this tendency).

So unbelievably effective are these purely human trends that a very good party trick can be performed as follows. First of all, write the number 7 on a piece of paper unobtrusively and lay it face down on the table. At some time later ask one of the company to promise to answer three very simple questions *immediately* you ask them, without any suspicious thoughts. These three questions are:

1. Three threes?
2. Name a colour.
3. Give me a number between 5 and 10.

When the number 7 comes up as the third answer you merely turn over the paper to show your hypnotic powers! About one person in 10 will answer 5, 8 or 9 for the third question. Such people are often of a 'cussed' nature. They will probably have answered 'purple' to question (2), whereas most people opt for red or blue. Invariably the unusual answer comes from one who did not truly empty their conscious mind before answering – a $\frac{1}{10}$-second delay is enough to allow this. The kind of 'logic' that goes on in the brain that conforms is this. 'Between' means 'somewhere in the middle'. That means 7 or 8. Eight is even, so non-random. It has to be 7.

Now computers work in binary digits, ones and zeros only. So surely we can do as well as the computer just by tossing a coin repeatedly. Again I tried as I was writing this. Starting with the coin head side up, I flicked it in the air 10 times and, writing '1' for head, '0' for tail, obtained:

0111010001

I then tossed it 10 times starting with the coin tail side up, and got:

1101111000

On the face of it, I was tossing randomly, 5 of each followed by a 6/4. Yet notice the grouping. In each sequence there is a tendency to restore the balance after about six or seven throws. In the first case, at one time it stood at 4 to 2, in the second 6 to 1. I watched the coin as it spun. It is possible to learn a technique for catching a spinning coin – with a desired result – from the sight of it in the air. Even blindfold and allowing it to fall on the floor is non-random. The method of flipping it would become biased after hundreds or thousands of tries, which is why the computer used the random flow of electrons in copper, as discussed in Chapter 8, p. 219.

In dealing with the very complex, such as the number of cells in the brain, the number of atoms in a cell, the number of grains of sand on the seashore and like quantities, mathematicians have devised a system of notation that avoids the use of the concept of infinity, for the latter is an admission of the inability of the human mind to contemplate such things, for example the probability of the existence of another planet exactly like the earth in every respect, including the number of humans, birds, insects, and so on, alive at any given instance. Sir Arthur Eddington once estimated the number of fundamental particles in the universe as 10^{89}, which written in decimal digits is 100 000 000 000 000 000 000 000 000 000 000 000 000 000 000 000 000 000 000 000 000 000 000 000 000 000 000 000 000.

I have written this in full, not to give an impression of how big it is, but of how small it looks, when you consider that there are about 300 000 000 000 000 000 000 molecules of gas in a cubic centimetre of air at ground level and about 2 000 000 000 000 000 000 cubic centimetres of air in the earth's atmosphere (assuming it to be of constant density for 10 miles and nothing outside of that). So the molecules of gas on earth alone amount to a 6 followed by 38 zeros or 6×10^{38}. But there is still a very long way to go to 10^{89}. There have to be about ten thousand million times the number of 'earths' in the universe that there are molecules on earth (taking into account the solid earth as well as its atmosphere), before you reach 10^{89}. At this point mathematicians said: 'Let us define a number bigger than 10^{89}, say 10^{100}.' (There is a lovely story of the mathematician who thought that there should be a special name for this number and asked his small son for suggestions. 'A googol' was the immediate reply. So a 'googol' it became!) Now, let us contemplate the number 10 raised to the power, a googol, i.e.

$$10^{\text{googol}} = 10^{10^{10}} = \text{a 'googolplex'.}$$

The number can never be written out in full in decimal digits, for there would be more zeros than there are fundamental particles in the universe!

Where is all this leading? Simply to the idea that it eliminates the need to use the word 'impossible'. For example, it would appear that as the result of our experiences it is impossible to pass one solid metal object through another in the sense that if two swords were crossed as in a duel they would always clash with each other and *never* pass through each other. Let us set this common-sense idea against our accepted concept of the structure of matter. The molecules of which the swords are made are themselves made up of atoms, and each atom consists of a nucleus, which we might represent as an orange in the centre of London, orbited by electrons (each the size of a pinhead), the nearest of which passes through Guildford and the furthest perhaps just north of Glasgow. The next atom has its nucleus in Cape Town! So it must surely be only a matter of getting all the nuclei and electrons in the two steel blades in the right *positions* for one set to pass entirely through the other set without disturbance? However persuasive this argument might appear, it has never been borne out in practice so far as we know. Now the purpose of the googol numbers is to state that all such impossible situations are not 'impossible', but the probability of their happening lies between a googol and a googolplex.*

Now all this fuss about 'randomness' and 'order' and 'disorder' was whipped into a frenzy by the thought that an electronic calculator that could work so fast and generate random numbers might be programmed to have a built-in critical faculty whereby it could pass judgement on a design of its own creation and could try millions of new designs in a matter of seconds. This would be

---

* An alternative way of expressing this idea became known as the 'megamon-key' technique. The suggestion was that if a million chimpanzees were put to work hammering on the keys of a million typewriters randomly, and successive generations of them continued for a million years, there were reasonable odds (in bookmakers' horse-racing terms) that one of them would have, over a period of several days, typed out faultlessly, and in order, the complete works of Shakespeare.

seen as comparable to the best of human inventions, even therefore to human thought processes themselves, and the science fiction writers cheered. Apparently a computer could become greater than its creator.

There were other facets of computers that were also comparable to living things. Even in the prototypes of very early computers which used thermionic valves, it was a tiresome task to look for the single valve that had failed in 10 000 or more. First, of course, you needed to *know* that this had occurred, and the method was simple. Program the machine to do the same sum twice, by different methods, and subtract the answers. (A very simple example would be $(27 \times 46) \div 19$, which could be done in the order stated there, and then repeated as $(27 \div 19) \times 46$.) If the result of the subtraction was other than zero, a light flashed, or a hooter sounded, (these days a pre-recorded voice might call out 'I'm sick'!). But then it was easy to devise a test program that effectively tested each of the 10 000 valves in turn, and the computer printed out the reference number of the valve that had failed. Although this was never actually incorporated, it was realised with the first machines of the early 1950s that it was then only a matter of straightforward engineering to arrange for the computer to replace its own burned-out valve to give it a process comparable with the self-healing processes that are present in every healthy living body. Our food needs a variety of constituents, proteins, carbohydrates, vitamins, and of course we need water and oxygen. The computer needs valves, resistors, capacitors, and a steady supply of electric power and cooling air. The comparison was a natural one.

By the late 1950s, faster computers were being made. The random soldering together of the components even of a prototype, which hitherto had looked like a bird's nest, had to be more organised. The location of the various master components such as the data stores had become critical. As my former tutor of those days put it: 'We are not yet limited by, but we have become conscious of, the transit time of information along a wire at the velocity of light, limiting the speed of computing'.

The way ahead was clear. Future computers had to be smaller, their components had to be smaller. Transistors replaced valves,

circuits became more 'ordered'. The structure of computer circuits photographed from a distance, slightly out of focus, began
to look like the cells of living tissue under a microscope. Were
we really so near to realising Frankenstein's dream? Fortunately,
no! Techniques were developed for 'printing' circuits on chips
of mica, then of printing the descendants of the transistor also,
until what had been a 6 foot by 2 foot Post Office rack of valves
and components of the early 1950s became a single chip less
than a square millimetre (0.006 square inch). Such components
have become *less* like living tissue, rather than more. We have a
whole order of magnitude yet to go.

A genius is often someone who can see further than others,
and at the time of the 1951 computer conference, the already
acknowledged mathematical genius Alan Turing made a calculation of the smallest machine that might be built whose only
purpose was to build other machines like itself. It must only obtain
energy from the sun, quarry its own minerals and refine them.
It must heal itself if it becomes damaged. He calculated that, on
1951 technology, its volume would be about one-half that of the
earth. Of course we have come a long way since then. On 1990
technology, perhaps it need be only the size of St Paul's
Cathedral; but in 1951 Turing contemplated that on a hot summer's day, that small black speck that is crawling across your
hand can do it! *That* is a fair statement of how far we have yet
to go, bearing in mind that the living black speck has no brain
nor central nervous system.

The advantages of the centralised nervous system

Whilst it was undoubtedly the emergence of the large-scale computer that triggered off the immodest thought that humans were
able to simulate their own brains, an interconnected system of
communications and action had been growing steadily for well
over half a century without anyone noticing that it had included
such complexity that when a computer was at last available, that
same computer became but a small part of the major system. I

refer, of course, to the electricity power supply system in the UK.

What is interesting is that from time to time in our history we have considered ourselves to be more than 'cast in the image of God' – we have seen ourselves as approaching a God-like quality. The Bible tells us of the Tower of Babel and the consequences that resulted from such arrogance. Present day science fiction tells us that the human mind almost demands the assumption of a God-like quality in itself. A TV series of the 1960s, with the title 'A for Andromeda', told a similar story of Babel-like disaster for those who thought that they had a recipe for creating human life, as the result of signals received from a distant constellation. There was much heated debate as to whether to proceed with such a project. The deciding sentence was to the effect that there was no real danger to humanity since if the system appeared to be getting out of hand, 'we can always pull the plug out'. What happened, of course, was that the first thing the machine did, as soon as it was large enough and capable of improving itself, was to make it impossible for any human to pull out its main plug!

Nature has a habit of doing such things. The Colorado beetle was a fairly rare beetle found only *in* Colorado, until Walter Raleigh showed civilised people the value of the potato. Since that time the beetle has been increasing in numbers and distribution to restore the balance whereby the potato plant is not grown to excess. By and large, the beetle is winning, despite modern chemistry.

An electronic computer of the 1950s was a 'glamour' object. It persuaded thousands of young people to take up electronics or computer programming as a profession in the belief that it pointed the way ahead in science and mathematics. Only a little of the 'shine' has worn off in the 1990s. Yet some of its by-products, at least, can now be seen as two steps forward and three back! In the bad old days, I used to be able to go into my local branch of the bank and ask what standing orders I had in force. The assistant would go to a filing cabinet and in a few minutes a list was forthcoming. One day recently I was told that I would have to wait a day 'because all the data were now stored in computer banks in Berkshire and there was a fault on the read-out'.

In the 1970s there appeared a printed notice in the centre

window marked 'Statements' to the effect that 'We regret that statements cannot be issued today owing to a computer fault'. The 'health' of the computer has become more important than that of any member of the bank's staff. In theory, shall we make better and better computers, until they *never* go wrong? That is not nature's way either. Mistakes are an important ingredient in progress.

Perhaps the biggest difference between the early computer and a brain is that the former dealt only in numbers as its ultimate product. A brain is there to control movements, habits, purposes and the survival of the species of which it is a part. When a computer is allowed to control something more than mere numbers it is said to be an 'on-line' computer, and takes a step nearer the living organism. The ensuing description of the functioning of the British electricity 'grid', in which almost every house and factory in the kingdom is interconnected with every power station, suggests that it might be the most complex project that has ever been attempted. The fact that the thinking behind it has become biological is perhaps only to be expected.

Figure 9.1 is a block diagram of the system broken down into a number of interconnected facilities. The block marked 'Power system' hanging on the bottom includes every power point in every house and factory served by the facility. Its complexity approaches that of a DNA molecule. The rest of the blocks together make up brain, sensors (tactile, auditory, visual, olfactory, etc.), genes (programmes for action as the result of signals from the sensors), nerves (transmission lines). The fact that the system cannot be simplified to the extent that a part of Fig. 9.1 cannot be enclosed by a dotted line and marked 'Brain' should tell us something about an animal's brain also, for both have been the result of evolution rather than of invention. Requirement alone has dictated design. Both can be seen to be successful. Success is built into the evolutionary programme.

Some explanations of the functions of the various blocks in Fig. 9.1 now follow. 'Generator scheduling' involves an hour-by-hour ordering of generation units on or off the system, just as an animal needs second-by-second (or an even shorter time scale) ordering of muscle power to meet the needs of physical movement. 'Econ-

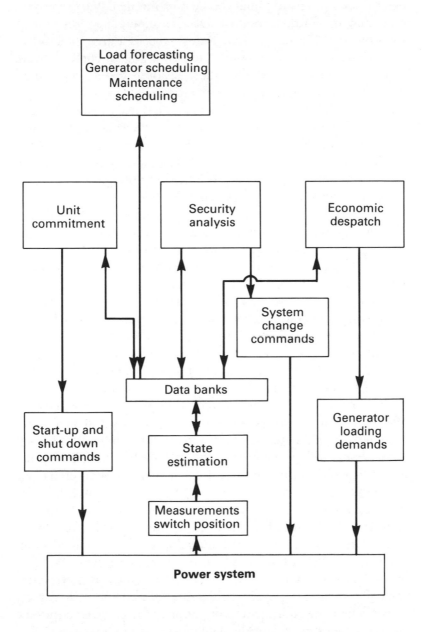

Fig. 9.1. Schematic diagram of an electricity supply system.

omic despatch' ensures that the cost of the demand is always a minimum in normal circumstances. In emergency, the advice from this unit can be ignored. Nature, too, is a great believer in conservation of muscle power. The power system control always anticipates the load needed on the basis of its experience, stored in the data banks. The power system has existed in an increasingly sophisticated form for over 50 years. It has seen many changes, emergencies, disasters. It has gone through several 'generations' of production, transmission and distribution techniques. As a result, its data banks are extensive. What animal does not use the experience gained by practice, or inherited through the genes, in order to forecast the probable muscular load?

Among the factors that allow power system demand prediction are such things as the regular daily habits of humans: time for starting work and for shut-down in factories, lunch and tea breaks. The demand varies from week to week throughout the year, not only as the result of obvious factors such as bank holidays and state visits, but as the result of climatic conditions. Weather forecasting is therefore a vital set of information bits. Changes in temperature, wind speed, humidity, cloud cover and visibility can all be seen on a chart of power requirement plotted against time. The ends of or commercial breaks in popular TV series when 20% of the nation 'gets up to put the kettle on' were one of the earliest eventualities to forecast! Who would have guessed, after only a cursory inspection of Fig. 9.1, that TV audience viewing figures would have a most important input to the data banks? The person with a clipboard who approached you in Market Street to ask what programmes you watched last night effectively caused you to act like a tiny part of the sensor of an animal (as it were, a single nerve ending), and your viewing information went into the Central Electricity Generating Board data bank.

The art of power demand forecasting has been refined to an extent that estimates are rarely in error by more than 3%. The average inaccuracy is less than $\pm 1\%$. I'm sure that a cricketer, footballer or tennis player would settle for *that* sort of accuracy, but of course the equivalent information in the game is required a thousand times more quickly. The power system forecasts always include a safety margin. Primitive power systems did not have this facility, but it is no stranger to the animal kingdom. We ourselves are only able to call on a fraction of our muscle power at any one time. The remainder is re-cycling and acting as stand-by in case of emergency. When someone sticks a pin into you in a surprise attack, you are allowed the use of the *whole* of the muscles necessary to take avoiding action. It is not a voluntary act. It is built into your system, virtually since you were an egg. If the restraining element breaks down, as it does in an epileptic during a fit, the whole of the muscles are available for voluntary use, and if used, the patient's muscles suffer afterwards. Whilst on the topic of voluntary/automatic animal activities, it has long troubled me that our creator did not trust us, so far as eyesight is concerned. Yet this is not true of all creatures. The domestic cat, for example, has control over its iris and, whereas in a dormant state only a slit of the lens is exposed, a slight movement in the grass, ahead of the cat, will cause it to open the iris fully, multiplying the available light by quite a large factor. Each of our eyes has an annular iris, as in cameras, and the easiest way to see it operate is to ask a friend (or look at yourself in a mirror) to cover one eye for a few seconds, then suddenly remove the cover whilst you watch the iris carefully. There is a time lag in the system so that you see the large black centre circle shrink to a smaller one, and if you are very observant you will see one overshoot – a phenomenon very common in all kinds of engineering practice, where any kind of control system is at work. Maybe we once had the facility to change the iris setting at will and we let it decay. Maybe we could develop the facility if we worked at it. Certainly we can lower our blood pressure or heart rate consciously with practice.

The 'Security analysis' facility of the power system assesses at all times the system response to a set of contingencies and provides a set of constraints (such as the ones just discussed in the case of animals), which must not be violated if the system is to remain in a secure state. The data banks are supplied from thousands of metering devices ('sensors' we would call them if we found them in an insect). The 'nerves' of the power system are usually telephone lines shared with the public for their more commonly known use. By comparison with the power flow lines, the telephone wires are thin, fragile lifelines, not unlike our animal nerves. The dual-purpose instrument was discussed at some length in earlier chapters, and found to be rare in engineering. But here, surely, is a case where engineers have used it as they do when they send data along power lines and, so far as we know, its counterpart in a nervous system has eluded us. Could this be an ideal opportunity for someone with the skill and knowledge to start looking?

The data banks of the power system cannot be discussed lightly as being so primitive in comparison with those of the brain of a bird or mammal that any suggestion of similarity is forcing the issue. Let us examine what happens to the information that arrives at the data banks. It is statistically inevitable that some unreliable information is fed to the data banks. Accordingly, the latter is connected to a 'State estimation' facility which is a mathematical algorithm* which provides a reliable data bank out of an unreliable set of information bits. This apparently impossible task *can* be performed whenever the number of bits of information is greater than that required for a complete load-flow solution.†

* An algorithm is a procedure for performing a complicated operation by carrying out a precisely determined sequence of simpler ones. The development of algorithms has introduced a hierarchical structure into human thought which has greatly increased its power. They exclude all personal judgement. A very simple example is digit-by-digit multiplication of large numbers (long multiplication), which only uses single-digit multiplication together with addition.

† This statement probably needs an example to substantiate it. It is convenient to use a science fiction setting for this purpose. The crew of a very advanced spaceship has invaded an enemy space-station and stolen their decoding apparatus which enables them to intercept all enemy messages. One of the crew is captured in the process. The raiding party intended to blow up the entire complex to disguise the fact that they had the decoder, but the explosives used were insufficient to do this. The enemy discovered what had been stolen and immediately began sending out messages about torturing the captive with a view to luring the invaders back to them so that they could destroy them, but the fact that they gave incorrect information about the captive was of no consequence. The fact that they were signalling on the topic at all told the invaders that their enemies now knew that they had stolen the decoder and therefore all future pick-up had a high probability of being diametrically opposite to the facts. Something useful came out of a pack of lies. An example as complicated as this is needed to illustrate the phenomenon. How much more complicated are the problems of forecasting a huge power network's requirements?

The more redundant data there is, the greater is the reliability. (Let me remind you again of the baby in Chapter 5 who *saw* a teething ring, *felt* it with its fingers, yet still put it to its tongue to collect more 'redundant' information.)

I need hardly add that the block marked 'State estimation' in Fig. 9.1 can, in the power complex, be a human being, and in this role such a creature is very often superb. A skilled and experienced individual can make remarkably accurate adjustments on the strength of a grossly corrupted set of information bits, provided there are enough of them. But to make a *machine* capable of such a feat represents an entirely new view of science. In the 'old science', every practitioner toiled for a lifetime to *perfect* their designs, to make measuring devices more and more 'correct' and 'infallible'. This leads essentially towards more perfect shapes, the shapes of Euclid.

This is not nature's way. Of all the leaves on a tree, very few could be described as 'perfect' or even 'typical'. For similar reasons there is no 'average' English man or woman. Nature accepts, as now do British Grid System designers, that mistakes are inevitable, they are part of the system, one might almost add 'an essential part'. The acceptance of mistakes as a *prima facie* ingredient, and the method of dealing with that ingredient lead naturally towards the complex, the tree with no two identical leaves, the forked lightning that is never straight. To ask for perfection *plus* the ability to eradicate mistakes is like asking for order and chaos simultaneously.

In the power system, one of the factors that influences its design is that a reserve turbo-generator that needs to be started 'from cold' may require 6 to 8 hours of preparation before it can become a useful part of the system. A 'hot unit' can be introduced in less than 3% of that time and be fully loaded in about 6% of it. The cold stand-by unit almost resembles one facet of the ability of living tissue to rush aid to a wounded section. The purpose of the unit commitment activity is to schedule the start-up and shut-down of units that are to be spinning and 'hot' without undue loss of economy. It resembles an animal's facility to make up a considerable blood loss in only a few hours. If the reserve of spinning generators becomes inadequate, the effect is to slow down the entire system, a result not unlike the behaviour of a sick animal.

The operating conditions in a power system are characteristic in terms of three operating states: normal, emergency and restorative. In the normal state the demands on the system can be met without violating the operating limits of any of the system's component parts. Severe disturbances such as the total loss of a generator, a transmission line blown down or a direct short-circuit on the line can be cushioned so well that the system may still settle down to a new normal state. It may, of course, be driven into an emergency state or beyond into a restorative state. In the emergency state virtually all of the demand is satisfied but with simultaneous violation of the operating limits in one or more system components. Corrective action is required to prevent or limit the damage to the overloaded components that would result. If the fault is intensely severe but localised, a part of the corrective action is to isolate, by opening circuit-breakers, the affected part (anaesthetise it). The removal of a piece of equipment in this manner will place extra load on other equipment, but this is not excessive in a well-developed system and may be tolerated without further damage until the corrective action has re-scheduled the pattern of the power flow to return the system from the emergency to the normal state. On rare occasions the isolation of a component may, because of its critical position in the system, cause further severe damage, and the isolation of successive parts escalates to the point of system breakdown. Now, neither load demands nor component limits are being met and the system is driven into the restorative state. A major contributory factor to such a catastrophe is the loss of stability amongst interconnected components. 'System stability analysis' is a mathematical procedure which studies the system beforehand and decides the corrective actions needed to retain stability whilst repair and replacement are carried out.

Even the normal state can be classified as 'secure' or 'insecure'. The system is secure if it can ride a major setback to one component without going into the emergency. 'Contingency analysis' consists of simulation by the computer of the occurrence of each possible failure in a list of components. The time taken to run through the list is proportional to the number of items in the list, and as it is desirable to study as few as possible, the data banks are asked for the benefit of experience as to the most likely danger spots. Contingency analysis is nothing more than a repetitive solution of hypothetical load–flow problems executed as many times as there are suspect items in the list. The results indicate whether the system is operating in a secure or insecure state. In the latter event, a calculation of the preventive control actions necessary to restore the control state follows immediately and the actions are taken. Fast and accurate automatic actions of this kind are a feature of the modern system.

That this detailed comparison is not based on an extension of real-life power system practice, operative only in the mind of the author just to make a point, may be evidenced by reading a text-book on 'electric energy' recently completed by a colleague and myself in our professional capacities.[1] There you will find many of the sentences of my worthy colleague that appear in the foregoing pages. Those who would seek a deeper understanding of the more primitive animals' thought processes could do worse than read Chapters 8 and 9 in the reference just quoted.

Are there any facets of the power system that might even teach us more about the central nervous system of living creatures than we have learned from conventional studies of anatomy and physiology? Surely, yes, even apart from the dual-purpose action of telephone lines already mentioned. Is it not almost a certainty that animals take in weather-forecasting data? Are all these 'old farmers' tales' to be disbelieved, for they will tell us with certainty that cows lie down before a rain shower to reserve a dry patch of grass. This is done not as the first drops begin to fall, but many minutes previously. As a boy I lived in the country. My father was a farmer. There was a pond near our house. Once or twice each summer a heron would visit the pond. (Herons are rare in the Fylde area of Lancashire.) Where it came from we never knew. But whenever it came, my father told me, it would rain heavily before nightfall. In my experience, it always did, even though the bird sometimes arrived on what promised to be a most glorious summer day.

This is more than weather forecasting, but like the latter subject it opens up the whole question of whether there are senses other than the traditional five known to us – as if I had not already raised this topic in earlier chapters! The power system can be seen as the analogue of several, quite different faculties of a well-developed animal.

The symptoms of a nervous breakdown are directly comparable to the conditions which obtain when a fault escalates until instability occurs, beyond which the whole system slows down and, if unchecked, finally comes to rest. The definition of a clinical nervous breakdown is when all muscular activity ceases. The mind goes on thinking but the will to move has crumbled. The sensors function perfectly. You can hear all, see all and you suffer no kind of paralysis such as is present in other malfunctions. The machinery is not destroyed. There is no physical injury. The demand simply exceeds the capability and all the safety devices rush to the aid of the apparatus so that it is not permanently damaged.

The fact that it is not possible to allocate a specific section of Fig. 9.1 to be designated 'brain' deserves a little elaboration, too. It was believed earlier this century that specific parts of the brain contained storage units for specific pieces of memory and for programmes of facilities. It was nothing short of a breakthrough in brain surgery when it was discovered that the problem was nothing like so simple. Whilst admitting, as we must, that the complexity of the human brain is as much an order of magnitude greater than that of the power system as the power system's complexity is greater than that of a child's model train layout, both analogies can surely give just a glimpse, a mere shadow, of the more advanced system from which we might profit. The model train layout, in particular, is an excellent toy to set any child on the way ahead more rapidly than can formal school lessons. It is not necessary to teach science historically in chronological order, as we largely tend to do at present. A continuation of this practice can only result (at the present escalation rate of new knowledge) in facing the problem, well before the end of this century, as to whether:

(a) to keep children in higher education until they are 35 years old, or

(b)   to allow specialisation starting from the age of 10.

Present-day experience more than suggests that we cannot afford (a) financially and that (b) represents a return to the dark ages where children were sent into factories from a very early age, never to emerge until retirement. The protagonists of 'broader education' were in full cry for nearly two decades (1955–1975). *Somebody* will have to handle the complexity, and a broader education for everyone makes no kind of sense. The only solution must lie in modelling new complexities as soon as their function has been understood, so that their incorporation into toys and games makes them self-evident long before their formal education demands their appreciation. Calculus can be taught successfully to children of 7 years old if it is palatably presented. I know this because I have done it.

Another vital activity is that we *must* do something about our lack of communication. Without breathing a word of heresy by suggesting that we follow the Russian lead in the study of hypnosis, telepathy and similar 'forbidden fruit', the brain surgeon must have lunch with the power system engineer at least once a week, play golf with him each weekend. Lessons learned in a pleasurable situation have a habit of being retained far longer than those learned by book reading. One of the advantages of a university structure is that the Professor of Nuclear Engineering may meet the Reader in Prosthetic Dentistry at lunch whether they make an arrangement to do so or not. The profit motive is a disease of human society that is only caused by the equivalent of drinking alcohol to excess. Nature strives to achieve the profitable. Profit seeking as an end in itself as well as in its purest, academic form is a key to survival, but not when it completely dominates the scene, for such a situation has no room for compassion. It is an uncivilised jungle.

References

Chapter 1

1. Warwick, R. & Williams, P. (eds.) (1973). *Gray's Anatomy*, 35th edn. London: Longman.
2. Paturi, F. R. (1976). *Nature, Mother of Invention – The Engineering of Plant Life*. London: Thames and Hudson.

Chapter 2

1. Laithwaite, E. R. (1980). *Engineer Through the Looking Glass*. London: BBC Publications.
2. Eddington, A. (1920). *Space, Time and Gravitation*. Cambridge: Cambridge University Press.
3. Ridley, B. K. (1976). *Time, Space and Things*. Harmondsworth: Penguin.
4. Teller, E. & Latter, A. (1958). *Our Nuclear Future*. Great Meadows, New Jersey: S. G. Phillips.
5. Bronowski, J. (1976). *Ascent of Man*. London: Book Club Associates. Published originally by BBC Publications (London) in 1973.
6. Lenihan, J. M. A. (1969). The triumph of technology. *Philosophical Journal*, **6**(1), 12–18.
7. Sprague de Camp, L. (1977). *Ancient Engineers*. London: Tandem.

Chapter 3

1. Dingle, H. (1965). *The Special Theory of Relativity*, 4th edn, p. 22. Methuen Monograph. London: Methuen.
2. Medawar, P. (1969). *Induction and Intuition in Scientific Thought*.

1968 Jayne Lectures of the American Philosophical Society. London: Methuen.

3. A speculation touching electric conduction and the nature of matter (1844). In *Faraday as Discoverer* (1877). London: Longmans, Green and Company.

4. Laithwaite, E. R. (1965). The goodness of a machine. *Proceedings of the Institution of Electrical Engineers*, **112**, 538–41.

5. Laithwaite, E. R. (1973). Magnetic or electromagnetic? The great divide. *Electronics and Power*, **19**, 310–12.

6. Gloede, W. & Wunderlich, K. (1976). *Nature as Constructor*.

7. Weiss Fogh, T. (1964). Diffusion in insect wing muscle, the most active tissue known. *Journal of Experimental Biology*, **41**, 229–56.

Chapter 4

1. Buller. A. R. H. (1909). *Researches on Fungi*. London.

2. Schwenk, T. (1965). *Sensitive Chaos*. London: Rudolf Steiner Press.

3. Thompson, D'Arcy Wentworth (1961). *On Growth and Form*. Cambridge: Cambridge University Press.

4. Stevens, P. S. (1976). *Patterns in Nature*. London: Peregrine Books. (First published in the USA by Little, Brown & Co. in 1974.)

5. Gulko, A. G. (1978). A fluid basis for matter and energy. *Speculations in Science and Technology*, **1**(2), Part I, 83–90; Part II, 165–73. The journal is edited by William M. Honig and published by WAIT in Perth, Western Australia.

6. Edmonds, J. D. (1978). The moon clock, time dilation and dynamic vacuum. *Speculations in Science and Technology*, **1**(1), 22.

7. Taylor, E. S. (1974). *Dimensional Analysis for Engineers*. Oxford: Clarendon Press.

8. Gregory, R. L. & Gombrich, E. H. (1973). *Illusion in Nature and Art*. London: Duckworth.

9. Gardner, M. (1978). Mathematical games. *Scientific American*, **238**(6), 18.

10. Haeckel, E. H. (1974). *Art Forms in Nature*. New York: Dover Publications.

11. Haeckel, E. H. (1904). *Kunstformen der Natur*. Leipzig & Vienna: Verlag des Bibliographischen Instituts.

Chapter 5

1. Bragg, W. H. (1925). *Concerning the Nature of Things*. Six Lectures Delivered at the Royal Institution. London: Bell.
2. Jerome K. Jerome (1970). *Three Men in a Boat*. Harmondsworth: Penguin.
3. Walton, G. N. (1964). Facts and artefacts. *The Modern Churchman*, **VII**, 233–8.
4. Morris, Desmond (1976). *The Naked Ape*. London: Jonathan Cape.
5. *Guinness Book of Records*, 27th edn (1981). Guinness Superlatives.

Chapter 6

1. Cullwick, E. C. (1939). *The Fundamentals of Electromagnetism*. Cambridge: Cambridge University Press.
2. Gardner, M. (1978). Mathematical games. *Scientific American*, **238**(6), 21.
3. Walton, G. N. (1964). Facts and artefacts. *The Modern Churchman*, **VII**, 233–8.
4. Purse, Jill (1974). *The Mystic Spiral*. London: Thames and Hudson.
5. Stevens, P. S. (1976). *Patterns in Nature*. London: Peregrine Books. (First published in the USA by Little, Brown & Co. in 1974.)
6. Schwenk, T. (1965). *The Sensitive Chaos*. London: Rudolf Steiner Press.
7. Thompson, D'Arcy Wentworth (1917). *On Growth and Form*. First reprinted in 1961 by Cambridge University Press.
8. Leopold, L. B. & Langbein, W. B. (1962). The concept of entropy in landscape evolution. *US Geological Survey*, Professional Paper **500A**, A1–A20.
9. Horton, E. E. (1945). Erosion development of streams and their drainage basins: hydrophysical approach to quantitative morphology. *Bulletin of the Geological Society of America*, **56**, 275–370.
10. Strahler, A. N. (1952). Hypsometric (area–altitude) analysis of

erosional topology. *Bulletin of the Geological Society of America*, **63**: 1117–42.

11. Murray, C. D. (1926). The physiological principle of minimum work applied to the angle of branching of arteries. *Journal of General Physiology*, **9**, 835–41.
    Murray, C. D. (1927). A relationship between the circumference and weight in trees and its bearing on branching angles. *Journal of General Physiology*, **10**, 725–39.
12. Ulam, S. (1966). Patterns of growth of figures: mathematical aspects. In Kepes, G. (ed.) *Module, Proportion, Symmetry, Rhythm*, pp. 64–74. New York: Braziller.

Chapter 7

1. Laithwaite, E. R. (1960). A radiation theory of the assembling of moths. *Entomologist*, **93**(1165 and 1166) 113–17, 133–7.
2. Callahan, Phillip S. (1977). *Tuning in to Nature: Solar Energy, Infrared Radiation, and the Insect Communication System*. London: Routledge & Kegan Paul.
3. Taken from Northrop, E. (1945). *Riddles in Mathematics*. London: English Universities Press.
4. Moore, R. E. M. (1966). Order as an apparently spontaneous product of chaos. *IFMSA Intermedica*, Summer, p. 31.
5. Moore, R. E. M. (1964). Some factors governing ancient mosaic design. *Bulletin of the Institute of Classical Studies*, **11**, 87.
6. Moore, R. E. M. (1966). Pattern formation in aggregation of entities of varied shapes as seen in mosaics. *Nature (London)*, **209**, 128–32.
7. Moore, R. E. M. (1972). Order–disorder phenomena in aggregations of particles of sizes and shapes which vary randomly within limits. *Journal of Microscopy*, **95**, Part 2, 243–7.
8. Moore, R. E. M. (1966). See note (4) above.

Chapter 8

1. Brock, T. D. (1964). Knots in *Leucothrix mucor*. *Science*, **144**, 870–71.

2. Gardner, M. (1970). *The Ambidextrous Universe.* London: Pelican Books.
3. Lipson, H. S. (1980). A physicist looks at evolution. *Physics Bulletin*, **31**, 138.
4. Anonymous (1967). Big insect take-over. *ECC Institute News*, **XX**(77), 16. Published by Electric Construction Work Institute, Wolverhampton.
5. Ardrey, R. (1961). *African Genesis.* London: Collins.
6. Berg, Howard C. (1975). How bacteria swim. *Scientific American*, **233**(2), 36–44.
7. Marshall, A. T., Lewis, C. T. & Parry, G. (1974). Paraffin tubules secreted by the cuticle of an insect *Epipyrops anomala* (Epipyropidae: Lepidoptera). *Journal of Ultrastructure Research*, **47**, 41–60.
8. Paturi, Felix R. (1976). *Nature, Mother of Invention: the Engineering of Plant Life.* London: Thames and Hudson.
9. Anonymous (1970). Is this altogether cricket, cricket? *New Scientist*, **47**(713), 273.
10. Prozesky-Schulze, Le *et al.* (1975). Use of a self-made sound baffle by a tree cricket. *Nature*, **255**, 142–3.
11. Callahan, Philip S. (1977). *Tuning in to Nature: Solar Energy, Infrared Radiation and the Insect Communication System.* London: Routledge and Kegan Paul.
12. Gamow, R. I. & Harris, J. F. (1972). *Spectrum*, **9**(8), 36–42.
13. They quote here Greguss, P. (1968). Bioholography, a new model of information processing. *Nature*, **219**, 482. My only comment is 'New?'.
14. Lenihan, J. M. A. (1972). What is bio-engineering? *Contemporary Physics*, **13**(3), 295–309.

Chapter 9

1. Laithwaite, E. R. & Freris, L. L. (1980). *Electric Energy: Its Generation, Transmission and Use.* London: McGraw-Hill.

Index

Bold numbers indicate major mentions